01-HHE-20T

AF622677

# REMOVAL OF ENDOCRINE DISRUPTING COMPOUNDS IN WATER RECLAMATION PROCESSES

by:

**Jörg E. Drewes**
Colorado School of Mines

**Jocelyn D.C. Hemming**

**James J. Schauer**

**William C. Sonzogni**
Wisconsin State Laboratory of Hygiene

*2006*

WERF

Awwa Research Foundation
Advancing the Science of Water®

NWRI

WATEREUSE

California Urban Water Agencies

The Water Environment Research Foundation, a not-for-profit organization, funds and manages water quality research for its subscribers through a diverse public-private partnership between municipal utilities, corporations, academia, industry, and the federal government. WERF subscribers include municipal and regional water and wastewater utilities, industrial corporations, environmental engineering firms, and others that share a commitment to cost-effective water quality solutions. WERF is dedicated to advancing science and technology addressing water quality issues as they impact water resources, the atmosphere, the lands, and quality of life.

For more information, contact:
Water Environment Research Foundation
635 Slaters Lane, Suite 300
Alexandria, VA 22314-1177
Tel: (703) 684-2470 Fax: (703) 299-0742
www.werf.org werf@werf.org

This report was co-published by the following organization. For nonsubscriber sales information, contact:

IWA Publishing
Alliance House, 12 Caxton Street
London SW1H 0QS, United Kingdom
Tel: +44 (0) 20 7654 5500
Fax: +44 (0) 20 7654 5555
www.iwapublishing.com
publications@iwap.co.uk


Library of Congress Catalog Card Number: 2006922100
IWAP ISBN: 978-1-84339-758



Wisconsin State Laboratory of Hygiene, Colorado School of Mines



The research on which this report is based was developed, in part, by the United States Environmental Protection Agency (EPA) through Cooperative Agreement No. CR827345-01 with the Water Environment Research Foundation (WERF). However, the views expressed in this document are solely those of Wisconsin State Laboratory of Hygiene and the Colorado School of Mines and neither EPA, WERF, AwwaRF, National Water Research Institute, WateReuse Foundation, nor California Urban Water Agencies endorses any products or commercial services mentioned in this publication. This report is a publication of WERF, not EPA. Funds awarded under the Cooperative Agreement cited above were not used for editorial services, reproduction, printing, or distribution.

# ACKNOWLEDGMENTS

We thank the facilities that agreed to participate in this study. Their willingness to provide information and samples was of obvious importance to the success of this project. The researchers would like to thank Drs. Ana Soto and Carlos Sonnenschein for the generous gift of the MCF-7 cells, and Janine Calabro for technical assistance with the assays. Dr. Kevin Gaido provided the transfected yeast assay, and Susan Ross provided advice on the assays. Also appreciated is the advice of Sigmund Deigitz, Joe Tietge, Jackson Gross, and Matt Meyer, who shared their expertise with the *Xenopus* assay.

## Report Preparation

**Principal Investigators:**
Jörg E. Drewes, Ph.D.
*Colorado School of Mines*

Jocelyn D.C. Hemming, Ph.D.
James J. Schauer, Ph.D., P.E.
William C. Sonzogni, Ph.D.
*Wisconsin State Laboratory of Hygiene*

**Project Team:**
Miel A.E. Barman
Mark G. Mieritz
Ryan T. Pieters
*Wisconsin State Laboratory of Hygiene*

Peter Fox, Ph.D.
*Arizona State University*

Joseph Gully, M.S.
*County Sanitation Districts of Los Angeles County*

Sarah J. Ladenburger, M.S.
Jessica Mansell, M.S.
*Colorado School of Mines*

## Project Subcommittee

Jim Crook, Ph.D., P.E., Chair
*Environmental Engineering Consultant*

Barnes Bierck, Ph.D., P.E.
*Environmental Engineering Consultant*

Michel Gibert, Ph.D.
*VEOLIA Environment*

Louis Guillette, Ph.D.
*University of Florida*

David Holbrook, Ph.D., P.E.
*National Institute of Standards and Technology*

Jerry Ongerth
*East Bay Municipal Utility District*

Shane Snyder, Ph.D.
*Southern Nevada Water Authority*

## Water Environment Research Foundation Staff

**Director of Research:** Daniel M. Woltering, Ph.D.
**Program Director:** Jami Montgomery
**Program Manager:** Jane M. Casteline

# ABSTRACT AND BENEFITS


**Abstract:**

As water demands increase, so do calls to use more treated wastewater to augment nonpotable and potable water supplies; yet people are concerned that recycled wastewater may contain contaminants harmful to human health or the aquatic environment. Contaminants of concern include endocrine disrupting compounds (EDCs), compounds that can interfere with the proper functioning of hormone systems. The goal of this research was to develop approaches combining bioassays with chemical analysis to study removal of EDCs by different reclamation treatment processes. Hydraulic corresponding composite samples were collected for individual unit operations and analyzed using HPLC-ELISA, GC-NCI-MS, and bioassays. Whereas estrogenic activity was accounted for by chemical components of primary effluents, more androgenic activity was found than could be explained by testosterone concentrations. Generally, secondary treatment removed EDCs very efficiently. For effluents in which estrogenic activity remained, it was mainly comprised of 17β-estradiol and estrone, whereas androgen activity was caused by remaining testosterone. Secondary treatment employing nitrification/denitrification more efficiently removed EDCs than facilities employing no nitrification. Membrane bioreactor treatment exhibited slightly better removal than conventional activated sludge. Powdered activated carbon, soil-aquifer treatment, and integrated membrane systems (microfiltration/reverse osmosis) efficiently removed endocrine disrupting activity. A tadpole metamorphosis assay was evaluated to assess potential thyroid disruption but proved more complicated than workable for a screening tool.


**Benefits:**

- Determines that two estrogenic bioassays (E-screen, yeast transfected with the estrogen receptor) agree reasonably well with chemical analyses, indicating that chemical contribution to estrogenic activity is well defined.
- Indicates that estrogenic activity in primary effluents is mainly comprised of natural hormones such as estriol and 17β-estradiol, whereas activity in secondary effluents is mainly caused by the presence of estrone and 17β-estradiol.
- Suggests that secondary treatment with SRT exceeding two days is efficient in significantly removing EDCs and biological activities.
- Determines that EDC load and removal correlate with BOD load and removal during secondary treatment.
- Shows that powdered activated carbon, subsurface treatment using soil-aquifer treatment, as well as integrated membrane systems (microfiltration/reverse osmosis) efficiently remove EDCs and biological activities to levels below the detection limit.


**Keywords:** Hormones, nonylphenol, octylphenol, estrogen, androgen, water reuse, bioassays, ELISA

# TABLE OF CONTENTS

# LIST OF TABLES

# LIST OF FIGURES

# LIST OF ACRONYMS

| | |
|---|---|
| APEC | Alkylphenol polyethoxycarboxylates |
| APEs | Alkylphenol polyethoxylates |
| AR | Androgen receptor |
| AR-yeast | Androgen receptor transfected yeast assay |
| AS | Activated sludge |
| A-screen | Androgen receptor transfected MCF-7 cell proliferation assay |
| BOD | Biochemical oxygen demand |
| BPA | Bisphenol A |
| CD | Charcoal dextran |
| CSM | Colorado School of Mines |
| DCM | Dichloromethane |
| DEHP | Di-(2-ethylhexyl) phthalate |
| DHT | Dihydrotestosterone |
| DI | Deionized water |
| DOC | Dissolved organic carbon |
| $D_p$ | Diffusion coefficient |
| EDCs | Endocrine disrupting compounds |
| EEq | Estrogen equivalents |
| EfOM | Effluent organic matter |
| EI | Electron impact ionization |
| ELISA | Enzyme-linked immunosorbent assay |
| ER-yeast | Estrogen receptor transfected yeast assay |
| E-screen | MCF-7 cell proliferation assay |
| F/M | Food to microorganism ratio |
| FBS | Fetal bovine serum |
| GAC | Granular activated carbon |
| GC-MS | Gas chromatography-mass spectrometry |
| GC-NCI-MS | Gas chromatography-negative chemical ionization-mass spectrometry |
| HFBAA | Heptafluorobutyric acid anhydride |
| HPLC | High performance liquid chromatography |
| HRT | Hydraulic retention time |
| $K_{ow}$ | Octanol-water partitioning coefficients |
| LOD | Limit of detection |
| MBR | Membrane bioreactor |
| MF | Microfiltration |
| MGD | Million gallons per day |
| MLSS | Mixed liquor suspended solids |
| MW | Molecular weight |
| MWCO | Molecular weight cutoff |
| NaCl | Sodium chloride |
| NCI | Negative chemical ionization |
| ND | Below detection |
| NOM | Natural organic matter |
| NP | Nonylphenol |
| NP1EO | Nonylphenol monoethoxylate |

| | |
|---|---|
| NP2EO | Nonylphenol diethoxylate |
| OD | Optical density |
| OEDC | Organisation for Economic Co-operation and Development |
| OP | Octylphenol |
| PAC | Powdered activated carbon |
| PCB | Polychlorinated biphenyls |
| PFBCl | Pentafluorobenzoyl chloride |
| $pK_a$ | Acid dissociation constant |
| ppb | Parts per billion |
| QA | Quality assurance |
| RO | Reverse osmosis |
| SAT | Soil-aquifer treatment |
| SRB | Sulphorhodamine B |
| SRT | Solid retention time |
| TCA | Trichloroacetic acid |
| TEq | Testosterone equivalents |
| TKN | Total Kjeldahl nitrogen |
| TOC | Total organic carbon |
| TSS | Total suspended solids |
| UV | Ultraviolet |
| WSLH | Wisconsin State Laboratory of Hygiene |

# EXECUTIVE SUMMARY

Results from recent research indicate that endocrine disrupting compounds (EDCs) are present in municipal wastewater effluent at concentrations capable of inducing feminization in fish. The feminization has been linked to exposure to compounds that mimic estrogen. However, thousands of compounds have the potential to interact with components of the endocrine system, altering the natural action of the hormone. This disruption can occur by several mechanisms, such as mimicking the hormone at a receptor (receptor agonist), blocking a hormone's normal action (receptor antagonist), or altering its synthesis, degradation or elimination. The estrogen receptor agonists are the most widely studied class of endocrine disruptors, and chemicals in this class include natural estrogens (17β-estradiol and metabolites estrone and estriol), synthetic estrogens (17α-ethinylestradiol used as active ingredient of contraceptives), plasticizers (bisphenol A and phthalates), and surfactants (alkylphenol degradation products nonylphenol and octylphenol). Although disruption in other hormone systems has not been as commonly observed, androgen and thyroid disrupting compounds may also be present.

Conventional wastewater treatment facilities are not specifically designed to remove EDCs, and, as reported in previous studies, the degree with which they are removed varies from nearly complete to very little during primary and secondary treatment. It is not clear whether observed variations of EDC concentrations and activities are a result of single unit processes, operational parameters, or sequences of processes employed during water reclamation. Also, there is a need to assess the removal of endocrine disrupting activity during advanced unit operations such as activated carbon, membrane treatment, UV disinfection, and soil-aquifer treatment (SAT). At a time when water reclamation is coming under more scrutiny as a way to conserve water resources, the purpose of this project was to improve the current understanding of removal of endocrine disrupting compounds and endocrine disrupting activities during water reclamation processes. Specific objectives of this study were to:

- Evaluate analytical tools, both chemical and biological, for quantifying endocrine disrupting compounds in wastewater matrices.
- Assess conventional water reclamation treatment trains to determine their ability to reduce concentrations of endocrine disrupting compounds and activity.
- Determine the degree to which operational parameters influence the final effluent concentrations of endocrine disrupting compounds and activity.
- Assess the removal of endocrine disrupting compounds by advanced treatment processes to determine their ability to reduce endocrine disrupting activity.

Bioassays are an important component of examining the presence and integrated effects of these EDCs. For example, degradation products or metabolites may also be biologically active even though the parent compound removal is indicated by chemical analysis. Additionally, the effect of chemical mixtures cannot always be inferred from their concentrations, and it is possible that not all active compounds have been identified. Thus, this project used a combination of biological and chemical analytical techniques.

Field investigations focused on water reclamation facilities located in California, Arizona, Colorado, Florida, Virginia, and Wisconsin, allowing an investigation of regional variations and the removal efficiencies of individual processes comprising a wide variety of process trains. Hydraulic corresponding composite samples were collected for individual unit operations. Secondary treatment was represented by activated sludge systems with hydraulic retention time (HRT) and solid retention time (SRT) as operational variables employing no nitrification, partial nitrification and nitrification/denitrification as well as chemical or biological phosphorus removal. In addition, a full-scale membrane bioreactor (MBR) was examined as well as three pilot-scale MBRs. Disinfection processes were represented by chlorination and UV disinfection. Laboratory investigations included exploring adsorption of EDCs onto powdered activated carbon (PAC) using spiked wastewater samples and testing of representative target compounds during treatment with high-pressure membranes under various operational conditions. In addition, soil-column studies to simulate SAT under different flow and redox conditions were conducted, and results were verified through followup full-scale studies at water reuse field sites in Arizona and Southern California.

Samples collected from the various sites and experiments were concentrated with solid phase extraction. Gas chromatography-negative chemical ionization-mass spectrometry (GC-NCI-MS) was used to quantify the target compounds, which included several steroid hormones (testosterone, 17β-estradiol, 17α-ethinylestradiol, estrone, estriol) and phenolic compounds (bisphenol A, 4-t-octylphenol, nonylphenol). Additionally, enzyme-linked immunosorbent assays after HPLC cleanup (HPLC-ELISA) were used to determine testosterone, 17β-estradiol and estriol concentrations. To evaluate estrogenic activity, the researchers used the MCF-7 cell proliferation assay (E-screen) and an estrogen receptor transfected yeast assay (ER-yeast). An androgen receptor (AR) transfected MCF-7 cell proliferation assay (A-screen) and an AR transfected yeast assay (AR-yeast) were used to assess androgen activity. The potency of the target chemicals in each of the bioassays was used to calculate the estrogen equivalents (EEq) or testosterone equivalents (TEq), which allowed for a comparison of chemical measurements and biological activity.

A tadpole metamorphosis assay was evaluated as a tool to indicate thyroid disruption using sodium perchlorate as a positive control as well as exposing tadpoles to a wastewater effluent. This in vivo test was of long duration (three weeks) and labor intensive. Although the results indicated differences in the stages of development reached during the three-week exposures, variability was fairly high and sensitivity to sodium perchlorate was lower than reported by others. The assay does show promise, especially if gene expression biomarkers could be developed to allow for reduced exposure times and greater sensitivity and specificity.

With respect to the first objective, evaluating the analytical tools, the research team found that working with the complex matrices in wastewater offered many analytical challenges. QA samples indicated blank contamination, especially with phenolic compounds. Recoveries of target compounds spiked into effluent organic matter ranged from 74% to 143% for the GC-NCI-MS, 30% to 63% for HPLC-ELISA, and 42% to 130% for the bioassays. A very strong relationship was found for the two estrogenic bioassays and GC-NCI-MS results, which indicated that all three analytical techniques are likely acceptable approaches with which to study the removal of estrogenic compounds and activity during wastewater treatment. Results for raw wastewater samples through GC-NCI-MS and HPLC-ELISA varied significantly but were in general agreement for treated wastewater samples. The androgen assays indicated higher levels of activity in the primary treated effluent than could be explained by the concentrations of testosterone, although the sensitivity of the AR-yeast assay was too poor to be able to assess activity in secondary treated wastewater.

In studies addressing the second objective, exploring the removal across conventional secondary treatment, secondary treatment can provide substantial removal of EDC compounds and activities. The removal appears most variable for estrone, nonylphenol, and 4-t-octylphenol. These compounds represent metabolites or degradation products and therefore have the potential to be formed during activated sludge treatment. It is noteworthy that 17α-ethynylestradiol was only present in one primary effluent out of seven full-scale facilities screened. While bioassays demonstrated a high potency for 17α-ethynylestradiol, findings of this study suggest a negligible contribution to the estrogenic activity in untreated and treated wastewater samples. While comparing the biological activity with the chemical analyses, the researchers found that difference in potency of target chemicals leads to different estimates of what components are responsible for the activity. Estriol has a greater potency in the E-screen than it does in the ER-yeast assay, whereas the alkylphenols are more potent in the ER-yeast assay than they are in the E-screen.

Therefore, the major active components of primary treated effluents for the E-screen were estriol (62%) and 17β-estradiol (21%), but the major active components for the ER-yeast were 17β-estradiol (42%) and alkylphenols (32%). For secondary treated effluents, estrogenic activity was mainly comprised of estrone (68%) and 17β-estradiol (22%) based on the E-screen potency factors and estrone (51%), 17β-estradiol (20%), and phenolic compounds (29%) based on the ER-yeast assay potency factors, respectively. Because of variability in responses of in vivo organisms, it is impossible to have one screening tool perfectly predictive for all organisms; however these in vitro tests appear to be reasonable estimates when compared with literature values. Although the combination of estrogenic compounds was well related to estrogenic activity, testosterone alone was not able to explain all the androgen activity in primary effluents. Other naturally occurring metabolites of testosterone (androsterone, 4-androstenedione, and 5α-androstanedione) would likely need to be quantified to find a better relationship.

In terms of the third objective, relating removal of EDCs with treatment parameters, findings suggest that the estrogenic activity in the influent is strongly correlated with the BOD load coming into a plant. Removal of biological activity correlated well with BOD removal, and high removal of individual EDCs and activity was achieved at solid retention times (SRT) exceeding two days. In the range of SRTs studied, no significant improvement in removal was observed between two and 10 days of SRT. A clear relationship was also lacking between removal and hydraulic retention time (HRT). However, one facility with the longest HRT, employing nitrification/denitrification and biological phosphorus removal, achieved the highest removal of EDCs among all plants investigated. This study did not consider facilities with a wide range of food to microorganism ratios (F/M ratio), but findings revealed a slight increase in removal efficiencies with decreasing F/M ratios. Lower F/M ratios indicate less available food for the community of microorganisms in the activated sludge system, which encourages the growth of a more diverse community of microorganisms with the ability to use less available carbon sources as food. Additional removal of steroid hormones was achieved during chlorination, whereas only a partial oxidation occurred for phenolic compounds. UV disinfection using low pressure, high intensity radiation had no effect on EDCs. Medium range UV radiation, however, achieved a partial degradation of phenolic compounds, whereas steroid hormone concentrations remained unchanged.

Advanced treatment processes, such as activated carbon, high-pressure membranes, and soil-aquifer treatment, represent viable barriers to further remove endocrine disrupting compounds as well as biological activities. Only small amounts of activated carbon (10 mg/L) were required to remove steroid hormones to below detection limits and significantly reduce phenolic compounds. Integrated membrane systems employing microfiltration followed by reverse osmosis were proven to remove EDCs and biological activity to no detection levels. Subsurface treatment, such as soil-aquifer treatment leading to groundwater recharge, reduced both EDCs and biological activities after very short travel times to below detection limits. These findings suggest that subsurface treatment is a strong barrier in removing biological activity from reclaimed water.

Findings from this study regarding the fate of EDCs during biological unit processes suggest that the main removal mechanism for EDCs and biological activity is biodegradation, with sorption to biosolids being important as well. Return flows from biosolids handling facilities (supernatent returns) can contribute significantly do the incoming load of EDCs to a wastewater treatment facility, suggesting that biosolid treatment is only partially successful in removing EDCs adsorbed onto biosolids. Future research needs to address the efficacy of different biosolids treatment processes, as well as the fate and transport of EDCs adsorbed onto land-applied biosolids.

The estrogenic in vitro bioassays proved to be robust tools for following changes in activity during wastewater treatment and were generally consistent with chemical measurements. Future work, however, could focus on the development of better in vitro tools for androgenic (especially improving sensitivity) and thyroid active compounds. With respect to gaining a better understanding of the impact of treated effluents on aquatic life, the ability to link physiological effects (intersex fish) with molecular endpoints, such as gene expression changes, will likely prove invaluable. Gene expression changes can occur with very short exposures, greatly reducing the expense of these studies. Gene array technology, where thousands of gene expression changes can be monitored in one experiment, shows great promise in the development of molecular biomarkers indicating exposures to EDCs.

CHAPTER 1.0

# INTRODUCTION

## 1.1 Introduction

Since the Clean Water Act was passed in 1974, wastewater in the United States has been required to undergo at least secondary treatment and disinfection before it is discharged to surface waters. This requirement has led to major reductions of the organic loading into receiving waters and vastly improved water quality. According to Johnson and Sumpter (2001), "This, together with other improvements, has permitted the return of a range of previously absent flora and fauna to many freshwater systems. Thus, environmental scientists have fauna to study and environmental regulators have something to protect again." So while many of the large problems of wastewater have been addressed, the focus has shifted to organic contaminants present at very low levels (μg/L and ng/L).

It was while studying fish downstream of a wastewater treatment plant that researchers discovered that endocrine disrupting compounds are present in municipal wastewater effluent at concentrations capable of inducing feminization in fish (Folmar et al., 1996; Jobling et al., 1998; Kavanagh et al., 2004). The feminization of fish is usually characterized by an increase in vitellogenin (a precursor to egg-yolk protein normally only found in females), alterations in sex hormone levels, development of the intersex condition (gonads containing both male and female specific tissues), or skewed sex ratios. The feminization has been linked to exposure to estrogenic compounds, both natural and synthetic (Desbrow et al., 1998). However, it has also been determined that thousands of compounds have the potential to interact with components of the endocrine system, altering a hormone's normal action. This disruption can occur by several mechanisms, such as mimicking the hormone at a receptor (receptor agonist), blocking a hormone's normal action (receptor antagonist), or altering its synthesis, degradation, or elimination. The best-studied example involves estrogen mimics. Although it is not surprising that natural and synthetic estrogens in receiving waters could interact with estrogen receptors in aquatic vertebrates, what was surprising was that a wide variety of other chemicals, including plasticizers and surfactants, also mimic estrogen at the estrogen receptor and cause estrogenic effects. Although disruption in other hormone systems has not been as commonly observed, androgen (male hormones such as testosterone) and thyroid disrupting compounds may also be problematic. For example, exposure of fish to pulp and paper mill effluents has resulted in masculinized fish (Larsson et al., 2000 and references therein).

Because of the growing realization that wastewater effluents may have a role in causing endocrine disruption, and at a time when water reclamation is coming under more scrutiny as a way to conserve water resources, it is imperative to assess the efficacy of water reclamation systems in removing potential endocrine disrupting compounds.

## 1.2 Background Information

### 1.2.1 Endocrine Disrupting Compounds (EDCs)

#### 1.2.1.1 Hormones

The steroidal sex hormones, estrogens (such as 17β-estradiol, estrone, and estriol) and testosterone, represent a class of hormonally active agents of particular interest because they are naturally excreted into the environment from human and animal sources as well as extensively used as pharmaceuticals (the synthetic hormone 17α-ethinylestradiol in birth control pills). Synthetic hormones are generally more stable in water than are natural hormones and generally have greater potency (Wegener et al., 1999; Metcalf et al., 2001; Snyder et al., 2001). Drewes and Shore (2001) estimate that a population of six million uses about 100 kg of estrogen per year. As a consequence, untreated domestic sewage effluent contains large concentrations of estrogen (40 to 300 ng/L) and testosterone (16 to 700 ng/L), which is highly variable depending on the source and dilution (Shore et al., 1993; Baronti et al., 2000).

17β-estradiol is one of three major human estrogens, the others being estriol and estrone. 17β-estradiol, along with other estrogens, directs the development of the female genotype in a developing embryo and at puberty. Human plasma estradiol levels can reach concentrations of 130 ng/L prior to ovulation (Lippert et al., 1999). 17β-estradiol is considered the most potent estrogenic hormone, and estrogenic activity in environmental samples is often reported in estradiol equivalents because all possible estrogenic compounds are not identified (Holbrook et al. 2002, Snyder et al. 2001). 17β-estradiol is synthesized from acetate, cholesterol, progesterone, and testosterone. It is produced mainly by the ovaries, and also the placenta and adrenal cortex in women, and produced in smaller quantities by the adrenal cortex and testes in males (Lauritzen, 1999). Circulating estradiol is rapidly converted to estrone in the liver, although the reaction is reversible. This metabolism is actually considered a means of regulating the cellular concentration of 17β-estradiol. Conversion of 17β-estradiol to estrone results in a decrease in overall strength of the estrogen activity. Estrone is a primary estrogenic component of several pharmaceutical supplements, including those containing conjugated and esterified estrogens. Estrone is further broken down (through a 16α-hydroxyestrone intermediate) into estriol, which also retains some estrogenicity (Lippert et al., 1999). Estriol is the major estrogen produced by the fetoplacental unit during pregnancy and is used as an indicator of the well-being of a fetus. Plasma levels of estriol can reach 110 μg/L in the final weeks of pregnancy. It exists in the blood in free and conjugated forms. It is also the most abundant estrogen in urine (Lippert et al., 1999; D'Ascenzo et al., 2003).

Testosterone is the primary androgen secreted into the bloodstream. In males, testosterone is produced in the Leydig cells of the testes by conversion of androstenedione and androstenediol to testosterone. In females, testosterone is produced from conversion of androstenedione, as well as production by the ovaries and adrenal glands. Testosterone is responsible for development and maintenance of the male secondary sex characteristics such as growth and maturation of the prostate and seminal vesicles. Presence in boys and absence in girls leads to the differentiation of internal and external genitalia. In most androgen target cells, testosterone is enzymatically converted to dihydrotestosterone (DHT), which is a more potent androgen than testosterone. Most circulating testosterone is converted by the liver into metabolites such as androsterone. Androsterone is then conjugated with glucuronic or sulfuric acid and is then excreted in the urine (Greenspan and Strewler 1997). Plasma levels of testosterone are approximately 0.5 μg/L in women and 6-10 μg/L in men.

Hormones are generally excreted from the body in the form of conjugates that are not biologically active. However, these conjugated hormones are readily deconjugated by enzymes present in the distribution system such that hormones are primarily in the unconjugated form by the time they reach the treatment plant (Baronti et al., 2000). Unconjugated hormones are found in wastewater influent at concentrations that would be expected based on excretion rates (Johnson and Sumpter, 2001). Belfroid et al. (1999) analyzed the raw influents of five wastewater treatment plants and found no detectable levels of the hormone conjugates investigated.

### 1.2.1.2 Phenolics and Phthalates

Several other chemical classes (alkylphenols, and some phthalates) have been shown to interfere with some endocrine-mediated processes in some systems, but evidence for effects on reproductive development in vivo is scarce. These compounds continue to be used in a variety of industrial applications worldwide (National Research Council, 1999). Alkylphenol polyethoxylates (APEs) are among the most widely used nonionic surfactants in a variety of industrial, agricultural, and household applications. Nonylphenol (NP), octylphenol (OP), and alkylphenol polyethoxycarboxylates (APEC) are biological metabolites of APEs. While APEs are not estrogenic, most of the alkylphenolic metabolites, including NP, OP, APECs, nonylphenol monoethoxylate (NP1EO) and nonylphenol diethoxylate (NP2EO), as well as their corresponding carboxylic acids (NP1EC and NP2EC), were reported to be estrogenic to fish at low microgram per liter concentrations (Jobling et al., 1996; Dussault et al., 2005). Nonylphenols have an estrogenic potency of approximately 1/10,000 that of 17β-estradiol (National Research Council, 1999). Phthalate esters are widely used in the plastics industry. In particular, di-(2-ethylhexyl) phthalate (DEHP) is produced in large quantities. Harris et al. (1997) conducted in vitro assays using a recombinant yeast screen and also tested for mitogenic effect on estrogen-responsive human breast cancer cells. A small number of commercially available phthalates showed extremely weak estrogenic activity (potencies ranged from approximately $10^6$ to $5 \times 10^7$ times less than 17β-estradiol). DEHP showed no estrogenic activity in these in vitro assays.

Another synthetic chemical that has measurable hormonal activity is bisphenol A (BPA). BPA is a chemical intermediate for numerous industrial products including polymers (polycarbonate plastics), epoxy resins, dyes, and flame retardants (tetra bromo bisphenol) (vom Saal, 2000). BPA is also used in dental sealants. BPA is continuously used and was one of the top 50 chemical products in the United States in 1995 (Kirschner, 1996). Nagel et al. (1997) reported that fetal exposure to environmentally relevant parts-per-billion (ppb) doses of BPA, in the range currently being consumed by people, can alter the adult reproductive system in mice. Its environmental relevance results from its low to moderate mobility in soils, which can potentially cause quality deterioration in artificial groundwater recharge systems. However, BPA does not persist and studies by Ike et al. (2000) revealed that BPA-degrading bacteria are relatively widely and abundantly distributed in biological wastewater treatment processes and the aquatic environment. Therefore, it is considered that BPA can be relatively easily removed in wastewater treatment processes and readily degraded in the aquatic environment. Belfroid et al. (2002) measured BPA in surface waters and concluded levels were not high enough to cause endocrine disruption in fish.

### 1.2.2 Occurrence of EDCs in Wastewater

Concentrations of some endocrine disrupting compounds commonly observed in primary and secondary treated wastewater are presented in Table 1-1. Generally, in secondary treated effluent, natural and synthetic hormones are present in the nanogram per liter range while alkyl phenols and bisphenol A occur in the microgram per liter range. It is clear that removal during secondary treatment is substantial; however, it is also clear that in some instances concentrations in the secondary effluent remain high enough to be of concern.

**Table 1-1. Occurrence of Target Compounds in Primary and Secondary Effluents.**

| Compound | Range in Primary Effluents (ng/L) | Range in Secondary Effluents (ng/L) |
|---|---|---|
| 17α-Ethynyl Estradiol | ND-13 [1,2,4,9,15,16] | ND – 7.5 [1,2,3,4,8,9,15,16] |
| 17β-Estradiol | ND-150 [1,2,5,6,9,11,12,13,15,16] | ND – 43 [1,2,3,4,5,7,11,12,13,15,16] |
| Estrone | 7.3-132 [1,2,5,6,9,11,15,16] | ND – 108 [2,3,4,5,7,9,11,15,16] |
| Estriol | ND – 263 [2,5,14] | ND – 18 [2,5,14] |
| Testosterone | 24-180 [11] | ND [11] |
| Nonylphenol | 1300 – 343,000 [11,12,13,14,16] | ND – 9100 [7,11,12, 13, 14,16] |
| 4-t-Octylphenol | 100 – 13,000 [11,13,16] | ND – 1300 [7,11,13,16] |
| Bisphenol A | 40-28,100 [10,11,12,13,16] | ND – 17,300 [7,10,11,12,13,16] |

ND = Below Detection. Sources: [1]Andersen et al., 2003; [2]Baronti et al., 2000; [3]Belfroid et al., 1999; [4]Carballa et al., 2004; [5]D'Ascenzo et al., 2003; [6]Desbrow et al., 1998; [7]Furuichi et al., 2004; [8]Huang & Sedlak, 2001; [9]Joss et al., 2004; [10]Lee & Pert, 2000; [11]Lee et al., 2004; [12]Matsui et al., 2000; [13]Nasu et al., 2001; [14]Solé et al., 2000; [15]Ternes et al., 1999a; [16]Vethaak et al., 2005

### 1.2.3 Removal of EDCs During Wastewater Treatment

Birkett and Lester (2003) provided a comprehensive description of the behavior of endocrine disrupting hormones in wastewater treatment processes. They listed four main removal pathways for endocrine disrupting compounds: adsorption onto suspended solids, aerobic and anaerobic biodegradation, chemical (abiotic) degradation via processes such as hydrolysis, and volatilization. In general research has supported the strongest role for adsorption and biodegradation for the removal of endocrine disrupting compounds.

#### 1.2.3.1 Sorption onto Biosolids

The octanol-water partitioning coefficients ($K_{ow}$) of hormones and phenols suggest that they should sorb to mixed liquor suspended solids (MLSS) before significant degradation occurs (Johnson and Sumpter, 2001; Holbrook et al., 2002). Indeed, sorption to biosolids has been found to be an important removal mechanism. Many studies have examined the removal of hormones by sorption to biosolids by comparing influent, effluent, and solids concentration of hormones in wastewater treatment facilities. For example, Holbrook et al. (2002) reported that the highest estrogenic activity was found in the digested biosolids. Takigami et al. (2000) found that hormones were adsorbed onto biosolids as the primary removal mechanism with the 17β-estradiol concentration of the biosolids 1,000 times greater than the secondary effluent concentration. Additionally, Clara et al. (2004) reported that bisphenol-A, 17β-estradiol, and 17α-ethinylestradiol were all highly adsorbed to both activated and inactivated sludge.

#### 1.2.3.2 Biodegradation

Biodegradation of 17β-estradiol, estrone, and estriol has also been demonstrated. Baronti et al. (2000) examined the concentrations of estriol, 17β-estradiol, and estrone at six wastewater treatment facilities. Effluent concentrations of estrone were reported at higher levels than influent concentrations for four of six sampling events, likely due to the oxidation of 17β-estradiol to estrone during the treatment process. Similarly, using activated sludge batch experiments, Ternes et al. (1999b) and Weber et al. (2005) found 17β-estradiol was removed with a concomitant increase in estrone. Estrone was biodegraded in both studies with longer incubation. The degradation of APEs during secondary treatment can also lead to the formation of the estrogenic nonyl- and octylphenols. In general, 17α-ethynlestradiol is less easily biodegraded than are the naturally excreted hormones (Ternes et al., 1999b; Layton et al., 2000; Shi et al., 2004; Weber et al., 2005).

#### 1.2.3.3 Photolytic Decay

The photolytic decay of hormones has been investigated by Gray and Sedlak (2003). The study reported a half-life of 3.5 days for 17β-estradiol when photolytic decay is the only removal mechanism. The study did not report high removal efficiencies in wetland field settings, indicating that there may be factors affecting the ability of hormones to undergo photolytic decay. Mansell et al. (2004) confirmed results reported by Gray and Sedlak (2003), finding that 17β-estradiol and testosterone were only removed by 10 percent over a 24-hour period.

#### 1.2.3.4 Relationship Between Removal of EDCs and Operational Parameters

Unfortunately, a fair amount of research reporting removal of estrogenic activity in wastewater treatment facilities fails to report operational parameters at the facilities studied. This omission has made it difficult to correlate the relationship between operational parameters and percent removal. Many studies have suggested that increased SRTs result in an improved removal of hormones in wastewater treatment facilities (Holbrook et al., 2002; Andersen et al., 2003). HRT has also been found to have positive correlation with hormone removal, and the longer the hormones are in contact with mixed liquor suspended solids (MLSS), the better the removal percentage of estrogenic hormones.

Strenn et al. (2003) reported that higher SRTs resulted in a greater percent removal of the hydrophobic compounds being studied and suggested that a minimum SRT is required for removal of EDCs. Lee et al. (2003) compared the removal of EDCs and SRT and HRT, attempting to fit the data with statistically relevant correlations, but were unable to find significant relationships. They did find that almost complete removal occurred at facilities with higher SRTs and HRTs. Kreuzinger et al. (2004) reported similar results and suggested that by achieving a low food to microorganism ratio of 0.2 – 0.3 kg $BOD_5$ / (kg TSS * day), an increase in micropollutant removal may be achieved. Clara et al. (2005) suggested that a critical SRT for the removal of the substance should be able to be determined. If the facility operates below the critical SRT, the removal will be minimal or primarily due to adsorption. The study determined that the critical SRT for bisphenol A, 17β-estradiol, estrone, and estriol was approximately 10 days. However, for 17α-ethinylestradiol they were unable to determine a critical SRT and other influences seem to be relevant in determining the removal of this compound.

### 1.2.3.5 Removal of EDCs During Advanced Treatment

The ability of advanced treatment technologies to remove endocrine disrupting compounds is an area of active research. Reverse osmosis, nanofiltration, membrane bioreactors and activated carbon are all being studied to determine whether removal can be improved for discharging into sensitive environments or water reuse.

***Reverse Osmosis and Nanofiltration Membrane.*** Nghiem et al. (2002) examined the adsorption of estrone to two different reverse osmosis membranes and found adsorption of estrone to one membrane was greatly affected by pH, whereas the adsorption of estrone to the other membrane type was unaffected by changes in pH. Nanofiltration membranes are porous and therefore employ a different removal mechanism than do reverse osmosis membranes. Nghiem et al. (2004) found that initially hormone adsorption to the membrane was the predominant removal mechanism of the compounds, whereas at later stages of filtration the compounds were rejected by size or charge exclusion.

***Membrane Bioreactors.*** Wintgens et al. (2002) investigated the ability of membrane bioreactor technology to remove both nonylphenol and bisphenol A and found that the removal of nonylphenol occurred primarily during the biological treatment by biodegradation or volatilization. The excess sludge contained a very small amount of both nonylphenol and bisphenol A. Snyder et al. (2004) found that two membrane bioreactors (Zenon and Mitsubishi) provided almost complete removal (detection limit of 1.0 ng/L) of estriol, estrone, 17α-ethinylestradiol, 17β-estradiol, testosterone, progesterone, and androstenedione. Only estrone was detected at 8.45 ng/L for one of the systems. Snyder et al. (2004) found limited removal of other pharmaceuticals and personal care products by the membrane bioreactor systems but acknowledged the possibility of total estrogenicity removal.

***Ultraviolet Disinfection.*** In a study investigating the degradation of EDCs during UV photolysis, Rosenfeldt and Linden (2004) found that because of a significant absorption minima around 250 nm, it was likely that monochromatic low-pressure lamps (which emit 254 nm) would not degrade EDCs. The researchers instead used polychromatic medium-pressure lamps in their experiments. The authors found that while UV treatment was effective at degrading bisphenol A, 17α-ethinylestradiol, and 17β-estradiol, the extended treatment times necessary would not be practical in many cases. Coleman et al. (2005) investigated ultraviolet photolysis and photocatalysis with titanium oxide of 17β-estradiol, estriol, and 17α-ethinylestradiol. They found all three compounds were degraded by photolysis alone, but they were degraded more quickly under photocatalysis. Rate constants for photolysis and photocatalysis were the most similar for 17β-estradiol with values of 0.174 and 0.134 per minute, respectively. 17α-ethinylestradiol showed the greatest ability to be degraded by both photolysis and photocatalysis, with rate constants of 0.231 and 0.195 per minute. Estriol was the least degraded by both photolysis (a rate constant of 0.156 per minute) and photocatalysis (a rate constant of 0.093 per minute).

***Activated Carbon Adsorption.*** Yoon et al. (2003) studied the effect of natural organic matter (NOM) on the adsorption of 17β-estradiol and 17α-ethinylestradiol. The most effective removal (greater than 90 percent) of the hormones was achieved in drinking water samples with low NOM concentrations that had been spiked with hormones. In full-scale activated carbon filtration application, the NOM may also block pores in the activated carbon, thus reducing the adsorption sites for the hormones (Yoon et al., 2003). The study also concluded that both increased contact time and increased dose were beneficial in increasing the removal of the hormones. Holbrook et al. (2001) did not see a decrease in estrogenic activity after using activated carbon in a secondary treated wastewater effluent. However, they did not report NOM concentrations, contact times, or doses. It is reasonable to assume that the samples Holbrook et al. (2001) analyzed had higher NOM concentrations than the drinking water used by Yoon et al. (2003).

In a study using activated carbon adsorption after MBR technology, Wintgens et al. (2002) found that granular activated carbon was capable of reducing nonylphenol and bisphenol A concentrations by 50 percent. The authors note that the relatively small reduction might be due to a rather short contact time in the granular activated carbon columns.

### 1.2.4 Analytical Techniques for EDCs

One of the challenges in this field is that analytical techniques must be very sensitive, as biological effects can occur from hormone exposure at low ng/L concentrations. The lowest observable effect concentration for the development of intersex in lab-exposed medaka was 10 ng/L and 0.1 ng/L for 17β-estradiol and 17α-ethynylestradiol, respectively (Metcalf et al., 2001). Additional problems can be due to the complicated matrices of wastewater influent and effluents. Thus, characterizing the limitations of the analytical techniques is very important in these analyses.

#### 1.2.4.1 Chemical Analyses

Gas Chromatography Mass Spectrometry (GC-MS) methods for the quantification of the steroids, hormones, and alkylphenols have been developed in the past to measure endocrine disrupting compounds in water samples (Ding and Wu, 2000; Kuch and Ballschmiter, 2001; Rudel et al., 1998; Ternes et al., 1999a,b; and Wegener et al., 1999). These methods employ electron impact ionization (EI), which typically yield limit of detections for these compounds in the higher ng/L concentration range. These detection limits may be suitable for many wastewaters after primary or secondary treatment but not be sufficiently sensitive to quantify all these compounds in the broad range of reclaimed wastewaters that were tested as part of this study. To this end, GC-MS techniques were used that employ negative chemical ionization (NCI), as previously described by Kuch and Ballschmitter (2001). The use of GC-NCI-MS allows detection of these chemicals at concentrations in the low ng/L range, which is in the range of detection limits of enzyme-linked immunosorbent assay detection (ELISA) but has significantly better specificity than ELISA (Huang and Sedlak, 2001).

ELISA is a simple, sensitive tool for detecting environmentally relevant concentrations of many pesticides and other chemicals of concern (Vanderlaan, 1991). Antibodies for many endocrine disrupting compounds have recently been developed and are now commercially available in ELISA kits. Currently, kits exist for many steroid hormones (17β-estradiol, estriol, testosterone), several nonylphenols, and bisphenol A. Chemical identification with GC-NCI-MS was used in conjunction with ELISA, as ELISA proved to be an inexpensive, novel tool useful to water utility personnel with little training in or access to more traditional chemical analysis methods. However, natural organic material present in wastewater effluents does interfere with commercially available ELISA kits; therefore a cleanup step using HPLC is necessary prior to analysis (Huang and Sedlak, 2001).

#### 1.2.4.2 Bioassays to Assess Potential Endocrine Disruption

Because endocrine disruption is a biological endpoint, bioassays are used to identify compounds with endocrine disrupting potential. The major concern regarding endocrine disrupting compounds in wastewater effluent has been estrogen receptor agonists. Receptor agonists are chemicals that bind with the hormone receptor and elicit a response. The researchers chose two types of in vitro tests for this study that quantify estrogen activity based on interactions with the estrogen receptor. The preponderance of data regarding the estrogenicity of effluents notwithstanding, androgens (such as testosterone) may also be a concern, as natural and synthetic forms of androgens are excreted by humans and other animals (Thomas et al., 2002; Soto et al., 2004; Svenson and Allard, 2004). Thus, in vitro assays capable of evaluating androgen receptor agonists were also used. In vitro assays are useful because they are based on known mechanisms, are generally relatively inexpensive, use small volumes of sample, are short in duration, and are often very sensitive. However they have drawbacks relative to in vivo systems because they have a lower relevance to whole organism health, do not account for differences in uptake, availability or metabolism, and only specific mechanisms are detected (Gray et al., 1997).

One estrogen receptor agonist assay is the MCF-7 cell proliferation assay, also known as the E-Screen. The assay relies on the ability of estrogen to induce proliferation in this breast cancer cell line. The assay was developed and used in the 1990s to identify several novel xenoestrogens. Advantages include that it is a mammalian cell line, with a real biological effect, and the assay is relatively easy. It is also one of the most sensitive assays, with an $EC_{50}$ between 2-3 ng/L (Zacharewski, 1997; Soto et al., 1998; Andersen et al., 1999). Theoretically, other mitotic agents can cause proliferation; therefore the mechansim is not definitely mediated by the estrogen receptor. However, no false positives have been identified (Zacharewski, 1997; Soto et al., 1998).

An analogous cell proliferation test for androgens (A-screen) has been developed using MCF-7 cells stably transfected with the pSG-hAR plasmid containing cloned human androgen receptor cDNA (Soto et al., 2004, Soto et al., 1998, Szelei et al., 1997). In these cells, inhibition of cell proliferation indicates androgen agonist activity. In these cells, the endogenous estrogen receptors are still present; thus the proliferative response still occurs. However, the androgen receptor is over-expressed in this cell line, and the binding of androgens to the androgen receptor causes induction of a protein (AS3) which triggers proliferative shutoff. Thus, the protein induced by the androgen receptor agonist overrides the estrogen proliferative effect. This assay was only recently developed and used for environmental samples (Soto et al., 2004), as there has been less interest in androgenic compounds.

Two transfected yeast assays were also used in the study. Receptor-mediated transcription assays rely on cells that are transfected with reporter gene constructs inducible by the receptor of interest. Reporter gene constructs are made up of a promoter region (a region of DNA that is responsive to the receptor), which is linked to a reporter gene. Reporter genes often include a gene encoding an enzyme that can be assayed spectrophotometrically (such as β-galactosidase) when the gene is expressed. Yeast cells do not have endogenous estrogen, thyroid or androgen receptors. Recombinant receptors have been cloned and transfected into yeast cells along with reporter gene constructs to create a yeast estrogen screen (ER-yeast) (Routledge and Sumpter, 1996; Gaido et al., 1997) and a yeast androgen screen (AR-yeast) (Gaido et al., 1997). Activity in these assays indicates that the action of the chemical is mediated through the receptor. These assays can be very sensitive, in the ng/L range (Zacharewsky, 1997; Andersen et al., 1999). They are easier to perform than cell proliferation assays and have a shorter incubation time. Very different results can occur from one strain to another, probably depending on the yeast subclone and specifics of the gene constructs. Some false negatives have been reported from yeast assays (EDSTAC, 1998). The false negatives could be due to differences in metabolism between yeast and mammalian cells and differences in permeability of compounds; yeast cells have a cell wall while mammalian cells do not. Additionally, agonists are not consistently discriminated from antagonists in these assays. For these reasons, the yeast screens are not recommended by EDSTAC (1998).

Determining whether any compounds are present that have the potential to disrupt the thyroid system is complicated by the myriad ways the thyroid system has been affected by chemicals (reviewed in Howdshell, 2002). Until recently, no environmental chemicals had been shown to act as thyroid receptor agonists (Kitamura et al., 2002). Because tadpole metamorphosis is driven by the thyroid system, tadpole metamorphosis assays are being developed for use as an indication of thyroid disruption (OECD, 2005, Degitz et al., 2005). This is an in vivo assay, with exposure of *Xenopus* tadpoles to the chemicals or effluents lasting greater than two weeks. The endpoints include rate of metamorphosis, histological changes in the thyroid glands and changes in the hormone levels. As this assay is still in the developmental stages, the research team evaluated the assay using sodium perchlorate as a positive control and exposed tadpoles to one secondary treated effluent.

## 1.3 Objectives

The goal of this project was to evaluate specific treatment operations for their ability to remove endocrine disrupting compounds. The research team used a multidisciplinary approach, with a team consisting of engineers, analytical chemists, and environmental toxicologists, to study removal of endocrine disrupting compounds at water reclamation facilities and at the bench- and pilot-scale. They focused on specific treatment processes, including activated sludge and UV disinfection, as well as advanced treatment methods such as membrane bioreactors, reverse osmosis, and soil-aquifer treatment. The laboratory-scale experiments studied activated carbon adsorption and rejection during reverse osmosis. The analytical techniques consisted of bioassays that indicate estrogenic (E-screen and ER-yeast) and androgenic responses (A-screen and AR-yeast) and two types of chemical analyses (ELISA and GC-NCI-MS). This information will be useful for optimizing removal of these compounds prior to reuse of the final effluent for various purposes or for discharge to the aquatic environment.

This study focused on endocrine disrupting activity changes in water reclamation systems in order to:

- Evaluate analytical tools, both chemical and biological, for quantifying endocrine disrupting compounds.
- Assess conventional water reclamation treatment trains to determine their ability to reduce concentrations of endocrine disrupting compounds and activity.
- Determine the degree to which operational parameters influence the final effluent concentrations of endocrine disrupting compounds and activity.
- Assess the removal of endocrine disrupting compounds by advanced treatment processes to determine their ability to reduce endocrine disrupting activity.

# CHAPTER 2.0

# FIELD INVESTIGATIONS

## 2.1 Introduction

Field investigations for this study were carried out at participating facilities located in six states. This approach allowed an investigation of both regional variations of effluent quality and the removal effectiveness of individual processes comprising an overall process train. The wastewater treatment processes studied included activated sludge (which is specified into no or partial nitrification, nitrification/denitrification, biological phosphorus removal, and chemical phosphorus removal); membrane bioreactors; ultraviolet and chlorine disinfection; integrated membrane systems (microfiltration/reverse osmosis); and natural treatment processes (soil-aquifer treatment and constructed wetlands).

## 2.2 Site Selection and Survey

The experimental design of this study considered the broad range of nonpotable and potable water reuse application currently employed throughout the U.S. Nonpotable reuse applications include landscape and agricultural irrigation, industrial process water, stream flow augmentation, as well as wildlife habitat maintenance. Indirect potable reuse includes surface spreading and subsurface spreading or direct injection projects. An understanding of EDC sources and their control in water reclamation facility watersheds provides source control options to complement treatment options. The service area of each participating utility was characterized during the initial phase of the study. A summary of the surveys for each utility can be found in Appendix A.

## 2.3 Sample Collection, Preparation, and Storage

Each facility was instructed to follow the guidelines presented in the document "Sample Collection for Bioassay and Chemical Analysis for Endocrine Disruption Activity," which is included in Appendix B. Efforts were made to reduce if not eliminate the use of plastics during the sampling process because they can either cause adsorptive losses of target compounds or leach endocrine disrupting compounds into the sample.

### 2.3.1 Sampling Equipment

The bottles used for sampling were pre-cleaned one-liter amber glass bottles (Eagle-Picher, Phoenix, Ariz.) provided by the Wisconsin State Laboratory of Hygiene (WSLH) to each of the facilities. Additional sampling equipment was rinsed with soap and water, hydrochloric or nitric acid, methanol, and finally rinsed with Type I water.

### 2.3.2 Sample Collection

Flow proportional, hydraulically corresponding, four-hour composite samples were taken at sampling locations throughout each of the water reclamation facilities. This ensured that samples were following the same slug of water through the facilities. Four-hour composite samples were collected at each sampling point by automatic composite samplers. The samplers were set to collect 250 milliliters of water every 10 minutes. For each sampling location, new tubing was installed, and the sampler was calibrated. The samples were collected in 6-L glass containers in coolers and kept cool during sampling. After collection, the samples were mixed thoroughly and adjusted to a pH of 2–3 with sulfuric acid (see Appendix C) and split into five 1-L sampling bottles. In some cases, where an automatic sampler was not available, a composite sample was assembled from grab samples that were taken at timed intervals and manually mixed. Some samples such as surface water and groundwater well samples were taken as one-time grab samples. Two liters of each sample were shipped to WSLH, and three liters were shipped to CSM via overnight courier on ice.

Before sampling, the proper operation of the plant was confirmed by operational parameters provided by the facility in the survey. Care was also taken not to sample the facilities within 48 hours following a rain event. In addition, all samples were collected on a Tuesday to eliminate effects from weekend flows.

### 2.3.3 Blanks

Two types of blanks were included in the study to account for possible contamination of the samples. For each facility, a travel blank was prepared at WSLH by filling a precleaned one-liter amber bottle with Type I water. The travel blanks were sent to the facilities with the sampling bottles and sent back to WSLH without being opened by the facility. Five 1-L Type I water sample blanks were also sent to each facility. The facilities were instructed to mimic the collection of the samples with the sample blanks. This included running the sampling blanks through the automatic sampler and acidifying the sampling blanks to a pH of 2–3 with sulfuric acid. Two blanks were shipped to both WSLH and CSM, along with the samples.

### 2.3.4 Shipment and Storage

Once the samples were collected and prepared, they were shipped overnight to WSLH and CSM in ice-packed coolers. Upon arrival, samples were logged and stored at 4°C pending further analysis. Within two days, samples were extracted. The extracts were stored at 4°C until analysis of extracts using bioassays, GC-NCI-MS, and HPLC-ELISA analyses were completed. Field and laboratory blanks were processed like field samples.

## 2.4 Secondary Treatment and Disinfection

Secondary treatment represented by activated sludge systems was studied at full-scale facilities. The types of secondary treatment represented in this study included no or partial nitrification, nitrification/denitrification, biological phosphorus removal, and chemical phosphorus removal. Table 2-1 summarizes the facility's location, main unit processes and sampling location, type of water reuse, number of samples taken, and sampling date.

Table 2-1. Summary of Full-Scale Activated Sludge Sampling at Participating Utilities.

| Facility | Unit Processes/ Sampling Location | Type of Water Reuse | Number of Samples Taken | Sampling Date |
|---|---|---|---|---|
| **Facility A** | *Activated sludge (nitrification/denitrification, biological P-removal); UV disinfection* | *Nonpotable reuse: Wildlife habitat maintenance/Stream flow augmentation* | 8 | *July, 2003* |
| | *Activated sludge (nitrification/denitrification, biological P-removal)* | | 2 | *July, 2004* |
| **Facility B** | *Surface water* <br> *Activated sludge (partial nitrification, chemical P-removal)* | *Nonpotable reuse: Wildlife habitat maintenance* | 3 | *September, 2003* |
| **Facility C** | *Activated sludge (no or partial nitrification)* | *Nonpotable reuse: Wildlife habitat maintenance* | 2 | *December, 2003* |
| **Facility D** | *Activated sludge (no or partial nitrification); tertiary filtration* | *Nonpotable reuse: Landscape and agricultural irrigation/ Industrial process water* | 3 | *March, 2004* |
| **Facility E** | *Activated sludge (nitrification/ denitrification); tertiary filtration UV disinfection* | *Indirect potable reuse: Surface spreading* | 4 | *October, 2003* |
| **Facility F** | *Activated sludge (nitrification/ denitrification); tertiary filtration* | *Nonpotable reuse: Agricultural and landscape irrigation* | 3 | *November, 2003* |
| **Facility G** | *Surface water* <br> *Activated sludge (nitrification/ denitrification)* | *Nonpotable reuse: Landscape and agricultural irrigation* | 3 | *December, 2003* |
| **Facility H** | *Surface water* <br> *Activated sludge (nitrification/ denitrification)* <br> *Activated sludge (no or partial nitrification)* <br> *Tertiary filtration* <br> *Disinfection (chlorination)* | *Nonpotable reuse: Landscape and agricultural irrigation Industrial process water* | 6 | *August, 2004* |

The facilities employing activated sludge treatment were all sampled between July 2003 and July 2004. A detailed discussion of the results from the activated sludge and disinfection samples can be found in Chapter 5.0 of this report. Facility A was sampled twice because the first round of samples could not be properly analyzed due to recovery issues in the GC-NCI-MS analysis. A detailed assessment of the samples taken in July 2003 at Facility A is provided in

Appendix D. The operational parameters for the facilities during sampling are summarized in Table 2-2.

Table 2-2. Summary of Operational Parameters of Full-Scale Activated Sludge Systems During Sampling.

| Facility | Design Flow ($m^3/d$) | Sampling Flow ($m^3/d$) | HRT (hours) | SRT (days) | MLSS (mg/L) | Organics Loading BOD (kg/day) | Influent BOD (mg/L) | Effluent BOD (mg/L) | Influent $NH_3$-N (mg/L) | Effluent $NH_3$-N (mg/L) | Effluent $NO_3$-N (mg/L) |
|---|---|---|---|---|---|---|---|---|---|---|---|
| A | 94,636 | 76,900 | 13.7 | 10 | 1,900 | 34,913 | 227 | 4.1 | 23.5 | 0.02 | 13.1 |
| B | 105,992 | 83,279 | 5.5 | 8.5 | 2,320 | 12,492 | 150 | 7 | 41 | 1.5 | NA |
| C | 136,275 | 104,894 | 1.2 | 1.7 | 2,180 | 18,042 | 172 | 4 | 30 | 33 | <0.20 |
| D | 236,588 | 237,270 | 2.3 | 10 | 3,660 | 40,810 | 260 | 2 | 34.5 | 2.4 | 6.64 |
| E | 68,137 | 37,854 | 5 | 10 | 2,000 | 7,041 | 186 | 4 | 25.5 | 0.13 | 6.59 |
| F | 87,064 | 68,932 | 6.8 | 4.4 | 2,229 | 9,857 | 143 | 4 | 28.5 | 4.74 | 2.05 |
| G | 238,481 | 238,481 | 7 | 7 | 2,000 | 62,005 | 172 | 2.2 | NA | 0.46 | 5.26 |
| H-1 | 632,164 | 480,747 | 4 | 7 | 3,000 | 79,323 | 140 | 8 | 24.8 | <0.1 | 6.1 |
| H-2 | 632,164 | 480,747 | 5 | 9 | 3,000 | 79,323 | 140 | 5 | 24.8 | <0.1 | 6.5 |

## 2.5 Advanced and Natural Treatment

A full-scale membrane bioreactor, a facility employing an integrated membranes system (microfiltration followed by reverse osmosis), two soil-aquifer treatment sites, and a constructed wetland represented advanced and natural treatment in this study. The facilities representing these types of treatment are summarized in Table 2-3.

**Table 2-3. Summary of Full-Scale Advanced and Natural Treatment Sampling at Participating Utilities.**

| *Facility* | *Unit Processes/ Sampling Location* | *Type of Water Reuse* | *Number of Samples* | *Sampling Date* |
|---|---|---|---|---|
| ***Facility E*** | *Tertiary treated effluent*<br>*Soil-aquifer treatment*<br>*(2 groundwater monitoring wells representing different travel times)* | *Indirect potable reuse:*<br><br>*Surface spreading* | 4 | *August, 2004* |
| ***Facility K*** | *Tertiary treated effluent*<br>*Soil-aquifer treatment*<br>*(2 groundwater monitoring wells representing different retention times)* | *Indirect potable reuse:*<br><br>*Surface spreading* | 3 | *March, 2005* |
| ***Facility I*** | *Membrane bioreactor*<br>*UV Disinfection* | *Nonpotable reuse:*<br><br>*Wildlife habitat maintenance* | 3 | *June, 2004* |
| ***Facility J*** | *Tertiary treated effluent*<br>*After microfiltration*<br>*After reverse osmosis (3-stage train)*<br>*(Feed, permeates, and concentrates of 1st, 2nd, and 3rd pass)* | *Indirect potable reuse:*<br>*Subsurface spreading/Direct injection*<br><br>*Nonpotable reuse:*<br>*Golf course irrigation* | 16 | *June, 2004* |
| ***Facility L*** | *Tertiary treated effluent*<br>*Constructed wetland effluent* | *Nonpotable reuse:*<br><br>*Stream flow augmentation* | 2 | *August, 2004* |

The facilities were all sampled between June 2004 and March 2005. A detailed discussion of the results from the membrane bioreactor samples can be found in Chapter 6.0 of this report. The results from the natural treatment processes (soil-aquifer treatment and constructed wetland) can be found in Appendix E. Chapter 7.0 contains a discussion of treatment efficiencies using high-pressure membranes, including the results from the full-scale integrated membrane system.

# CHAPTER 3.0

# ANALYTICAL METHODS

## 3.1 Chemicals and Materials

All chemicals used were of reagent grade or higher, purchased from J.T. Baker, Inc. (Philippsburg, N.J.), CDN Isotopes (Quebec, Canada) Sigma-Aldrich (St. Louis, Mo.), Eastman Organic Chemicals (Rochester, N.Y.), and Fisher Scientific (Fairlawn, N.J.).

## 3.2 Target Compounds

The target compounds selected for this study are steroid hormones (Figure 3-1), major degradation products of non-ionic surfactants, and bisphenol A (Figure 3-2). These target compounds are characterized as hydrophobic with log $K_{ow}$ (octanol-water distribution coefficient) above 2.4.

**17β-Estradiol**
Mw 220.4
Log $K_{ow}$ 4.0
CAS # [50-28-2]

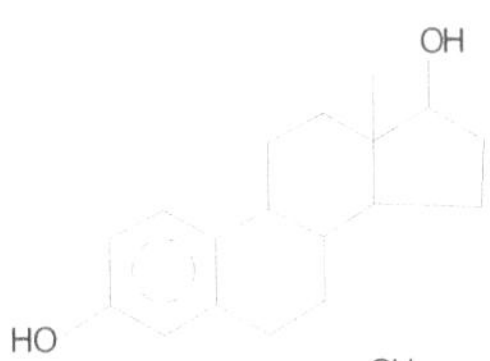


**17α-Ethinylestradiol**
Mw 272.4
Log $K_{ow}$ 4.5
CAS # [57-63-6]

**Estriol**
Mw 288.4
Log $K_{ow}$ 2.4
CAS # [50-27-1]

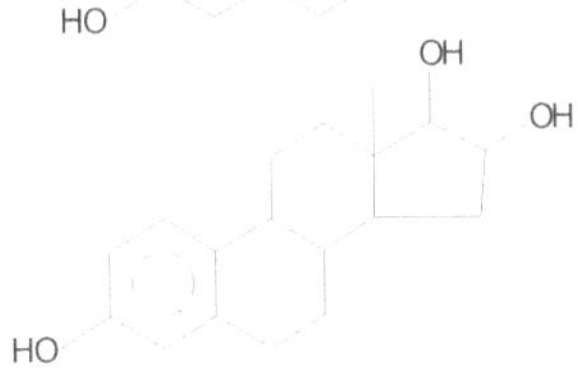


**Testosterone**
Mw 288.4
Log $K_{ow}$ 3.3
CAS # [58-22-0]

**Estrone**
Mw 270.4
Log $K_{ow}$ 3.1
CAS # [53-16-7]

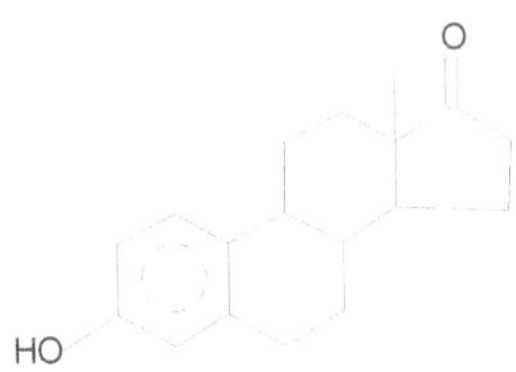


**Figure 3-1. Target Hormones.**

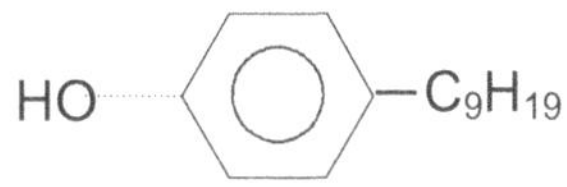


**4-Nonylphenol**
Mw 220.4
Log $K_{ow}$ 6.2
CAS # [104-40-5]

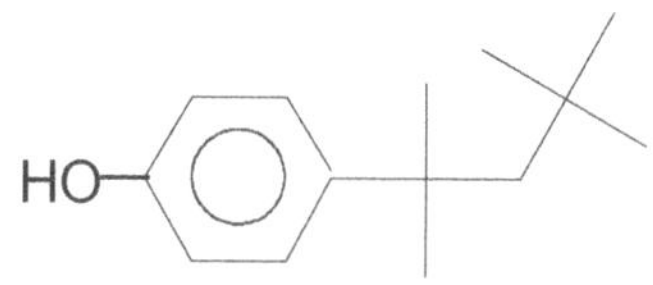


**4-(tert-Octyl)phenol**
Mw 206.3
Log $K_{ow}$ 5.7
CAS # [140-66-9]

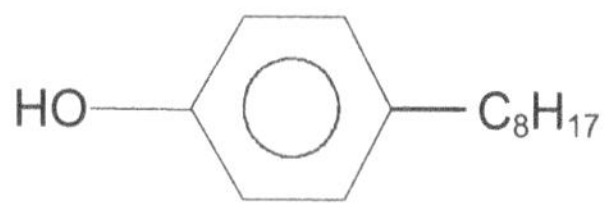


**4-Octylphenol**
Mw 206.3
Log $K_{ow}$ 5.7
CAS # [1806-26-4]

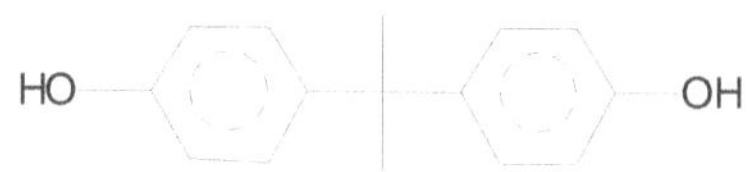


**Bisphenol A**
Mw 228.3
Log $K_{ow}$ 3.3
CAS # [80-05-7]

**Figure 3-2. Target Phenolics.**

## 3.3 Extraction Methods

### 3.3.1 Extraction for Gas Chromatography with Mass Spectroscopy and Bioassays

Sample concentration was achieved by filtering a 1-L sample through a series of two high-performance extraction disks (C18 and a SDB-XC disk 3M Empore™ St. Paul, Minn.). The C18 disk was cleaned with 10 mL of a 50/50 dichloromethane (DCM)/ethyl acetate solution and conditioned with 10 mL methanol, followed by a 20-mL Type I water rinse. The 1-L sample was filtered through the disk and then eluted, first with 5 mL of ethyl acetate, then 5 mL of a 50/50 mixture of ethyl acetate and DCM, followed by 5 mL DCM alone. The SDB disk was swelled and prewashed with acetone followed by isopropyl alcohol. The disk was then washed with 10 mL of DCM. Finally, the disk was conditioned with 10 mL of methanol followed by a 20-mL Type I water rinse. The sample previously run through the C18 disk was modified by adding 50 g muffled NaCl, and the pH was lowered to 2 with DCM-cleaned HCl. The sample was pulled through the disk, the disk was rinsed with 100-mL Type I water, and was eluted with 3 x 5-mL rinses of DCM. The extracts from each disk were combined and blown to near dryness with nitrogen. The extracts were transferred with several ethanol rinses to a calibrated 2-mL amber vial, dried to near dryness, and brought back up to 1.5 mL in ethanol. Samples were stored in a $4^{o}C$ cooler until use in the bioassay or GC-MS analysis.

### 3.3.2 Extraction for HPLC-ELISA

Samples were processed through a C18 solid phase extraction disk (Empore, St. Paul, Minn.), precleaned and conditioned with methanol. Samples with elevated dissolved organic carbon (DOC) concentrations were processed with a methanol-rinsed stainless steel pressure filtration system using medical grade nitrogen gas. Organic matter was extracted from the C-18 disks using a 40/60 methanol/water solution and was discarded. Two subsequent elutions with 90/10 methanol/water were collected. The extracts were dried completely either under nitrogen gas or in a 45°C oven. The dried extracts were resuspended in 1mL of a 15/85-acetonitrile/water solution.

## 3.4 Chemical Analyses

### 3.4.1 Gas Chromatography-Mass Spectrometry

The ethanol extracts were analyzed by two different gas chromatography-negative chemical ionization-mass spectrometry (GC-NCI-MS) methods to quantify the hormones and phenolic compounds. 17β-estradiol, estrone, 17α-ethinylestradiol, and the phenols (4-nonylphenol, 4-t-octylphenol, 4-octylphenol, and bisphenol A) were derivatized with pentafluorobenzoyl chloride (PFBCl) and analyzed by GC-NCI-MS (Kuch and Ballschmiter, 2001). Testosterone, estriol, 17β-estradiol, estrone, 17α-ethinylestradiol, and the phenols were derivatized with heptafluoric butyric acid anhydride (HFBAA) and analyzed by GC-NCI-MS (Table 3-1). The primary quantification for testosterone and estriol was based on the HFBAA derivatives. The primary quantification for the remaining chemicals was based on the PFBCl derivatives. Detection limits for the target hormones were based on the lowest standards and were in the range of 0.6-4.7 ng/L. Detection limits for the target phenolic compounds were typically limited by field blanks and laboratory blanks. Two times the standard deviation of blanks (50-106 ng/L) was used as the detection limit for the GC-NCI-MS analysis of the phenols. 17β-estradiol-$d_5$ and bisphenol A-$d_{16}$ were used as internal standards for the quantification of the hormones and phenolic compounds, respectively. The sample aliquots used for hormone analyses were spiked with both of the internal standards prior to derivatization.

The PFBCl derivatization was performed by evaporating the sample to dryness using high purity nitrogen gas and then resolvating the sample in 2 mL of high purity water. 50 µl of 2M KOH and 10 µl of 10 percent solution of PFBCl in toluene were then added to the resolvated sample. The derivative was extracted from the aqueous solution with hexane. The hexane extract was then analyzed by GC-NCI-MS, using methane as a reagent gas. The internal standard areas for each sample and standard were used to verify that the derivatization reaction was successful. Each batch of samples contained field blanks, laboratory blanks, and spiked samples for quality assurance and quality control purposes. The hormones, bisphenol A, 4-t-octylphenol, and 4-octylphenol were quantified using multipoint calibrations curves generated with dilution of authentic standards. Nonylphenol was quantified using multipoint calibration curves generated with a mixture of nonylphenols. The quantification of nonylphenol was based on an average response factor for the numerous isomers that elute at similar retention times.

Table 3-1. GC-MS Methods Used for Target Compounds.

| Compound | Derivatization Method | GCMS Analysis | Quantification Ions | Confirmation Ions |
|---|---|---|---|---|
| 17β-estradiol | PFBCl | GC-NCI-MS | 466 | 448 |
| | HFBAA | GC-NCI-MS | 427 | 197, 644, 467 |
| Estrone | PFBCl | GC-NCI-MS | 464 | 448 |
| | HFBAA | GC-NCI-MS | 197 | 174, 192, 160 |
| 17α-ethinylestradiol | PFBCl | GC-NCI-MS | 490 | 472, 474 |
| | HFBAA | GC-NCI-MS | 197 | 467, 644 |
| Testosterone | HFBAA | GC-NCI-MS | 483 | 197, 443 |
| Estriol | HFBAA | GC-NCI-MS | 679 | 197, 446, 639 |
| 4-tert-octyphenol | PFBCl | GC-NCI-MS | 400 | 382 |
| | HFBAA | GC-NCI-MS | 197 | 364, 177 |
| Nonylphenol | PFBCl | GC-NCI-MS | 414 | 396 |
| | HFBAA | GC-NCI-MS | 197 | 494, 177, 160 |
| 4-Octylphenol | PFBCl | GC-NCI-MS | 414 | 396 |
| | HFBAA | GC-NCI-MS | 197 | 364, 177 |
| Bisphenol A | PFBCl | GC-NCI-MS | 616 | 406 |
| | HFBAA | CG-NCI-MS | 197 | 423, 177 |

Note: 17β-estradiol, estrone, 17α-ethinyl estradiol, 4-t-octylphenol, nonylphenol, 4-octylphenol, and bisphenol A were quantified by both GC-MS derivatizations. The primary quantification was from the PFBCl and the HFBAA derivatization was used for confirmation.

The sample aliquots for HFBAA derivatization were also blown to dryness using high purity nitrogen after being spiked with the two internal standards. The sample was then resolvated with 100 µL of acetonitrile and 120 µL of HFBAA. After vortexing for 20 seconds, the mixture was heated for one hour to 60°C. The derivative reaction mixture was then re-evaporated to dryness using high purity nitrogen. The derivatized products were then solvated with n-hexane and analyzed by GC-NCI-MS under the same conditions as the PFBCl derivatives.

### 3.4.2 HPLC-ELISA

The analytical method employed for quantification of steroids using high pressure liquid chromatography coupled with enzyme immunosorbent assays (HPLC-ELISA) in water samples was adapted from Huang and Sedlak (2001). A Hewlett Packard Series 1050 HPLC, equipped with two exchangeable columns and an ultraviolet (UV) detector (at 220 nm), was used to isolate the fractions of samples that contain the compounds of interest. The first column used was a gel permeation/size exclusion column (Alltech) with a packing material size of 300Å. A standard mixture of hormones (estrone, estriol,17β-estradiol, testosterone) was injected into the column to determine the detention time of the compounds. The gel permeation column used a mobile phase of 85% water and 15% acetonitrile at a flow rate of 1 mL/min. To obtain optimal peak separation, a 200-µL sample volume was drawn into the injection needle at a rate of 300 µL/min and ejected from the needle into the column at a rate of 800 µL/min. (Figure 3-3). After two Type I water flushes, the sample was injected under the same conditions as the standards, and the fraction of the sample that corresponded to the detention times of the compounds of interest was collected and dried under a gentle stream of nitrogen. Once dried, the fractions were reconstituted in 50/50-acetonitrile/$H_2O$ solution for use on a reverse phase HPLC column.

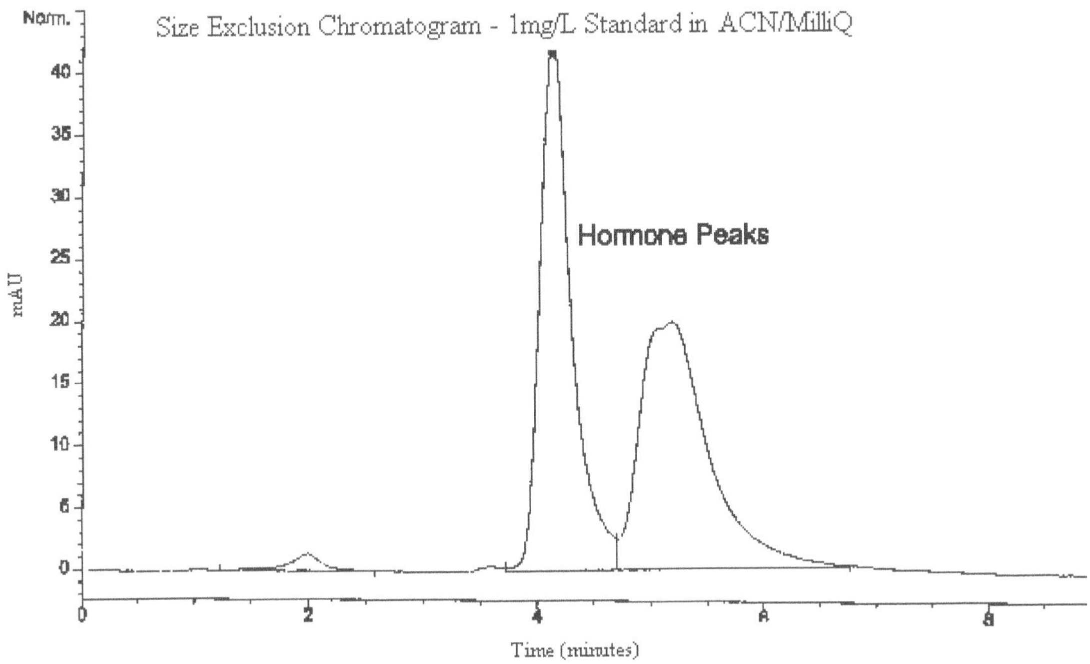


**Figure 3-3. Size Exclusion Chromatogram for Separating Organic Matter Background from Hormones.**

The reverse phase column used 5 μm particles, 9% carbon load. A standard mixture of 1 mg/L of each hormone in a 50/50-acetonitrile/$H_2O$ solution was injected into the column at a rate of 1 mL/min with a mobile phase of 55/45-acetonitrile/$H_2O$. The detention time and peak width of each hormone in the reverse phase column was determined by injecting individual standards. The mobile phase ratio was adjusted to achieve a minimum peak width. After flushing the injection needle with two injections of Type I water, each sample was then injected. A one-minute fraction prior to the estriol peak, the estriol peak, and a one-minute fraction after the estriol peak were collected. The estrone, 17β-estradiol, and testosterone peaks were collected as one three-minute fraction because these three compounds had similar retention times in the column (Figure 3-4). This one fraction is used for analysis of all three compounds on their respective ELISAs since the cross reactivities of a specific kit with other compounds is minimal. The fractions collected after reverse phase chromatography were completely dried under a gentle stream of nitrogen. The fractions were then resuspended in the buffer solution provided by the ELISA kit.

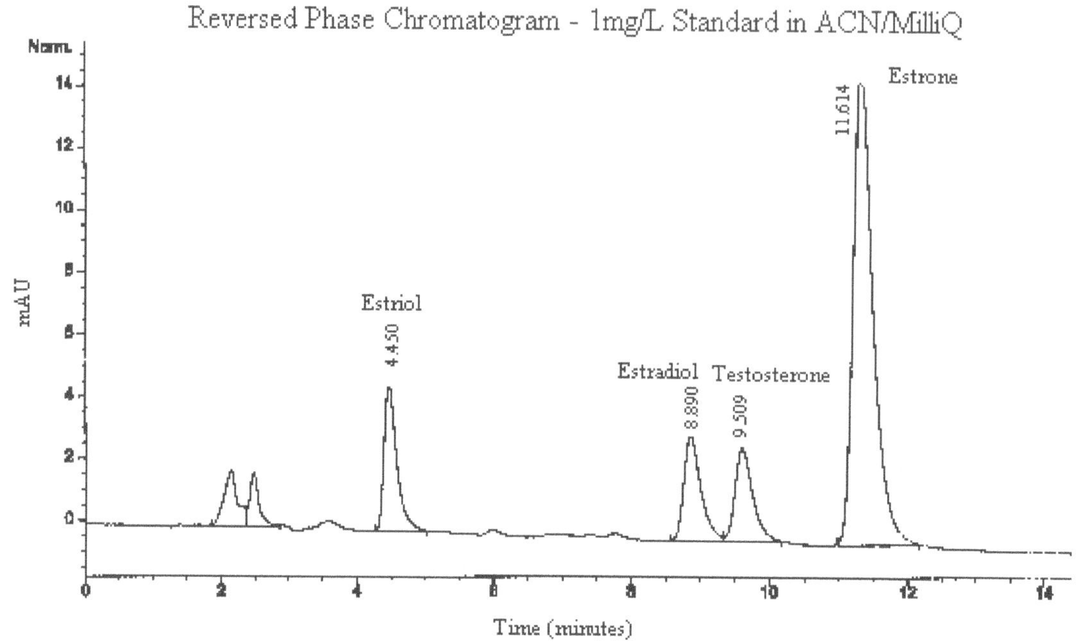


**Figure 3-4. Reverse Phase Chromatogram for Individual Target Compounds.**

Quantification of the steroids in the samples was performed using ELISA. 17β-estradiol, estriol, and testosterone ELISA kits were obtained from Cayman Chemicals (Ann Arbor, Mich.). The recoveries for all three target compounds through the extraction procedure and the HPLC cleanup were 95 ± 3%. Recoveries ranged from 45% to 95% when isolated compounds were quantified using ELISA. Recoveries of all compounds were verified using a standard addition of 30 ng/L added to a split sample prior to solid phase extraction. Laboratory blank samples using ultrapure water were processed along with each batch of field or laboratory samples. The detection limit of the analytical method for a 1-L sample for 17β-estradiol, estriol, and testosterone was calculated according to Standard Methods (AWWA, 1998) by considering three times the standard deviation of the blank samples. Limits of detection (LOD) for 17β-estradiol, estriol, and testosterone were 0.4, 0.6, and 0.5 ng/L for a 1-L sample, respectively.

## 3.5 Bioassays

### 3.5.1 Cell Proliferation Assays (E-screen and A-screen)

#### 3.5.1.1 Cell Culturing

The E-screen uses MCF-7 cells, a breast cancer cell line that proliferates in response to estrogenic compounds. The A-screen uses MCF7-AR1 cells, which are stably transfected MCF-7 cells that express the human androgen receptor. The MCF7-AR1 cells indicate androgen activity by decreasing proliferation of cells exposed to estradiol. The cells were obtained from Drs. Sonnenschein and Soto at Tufts University (Boston, Mass.). The MCF-7 and MCF7-AR1 cells were cultured in Dulbecco's modified eagle's medium (ICN Biomedicals, Aurora, Ohio) with 5% fetal bovine serum (FBS) (Hyclone Laboratories, Logan, Utah) at 37°C and 6.5% $CO_2$ in 75 $cm^2$ tissue culture flasks. Media was changed every two to three days, and the cells were subcultured every seven days. Stocks of MCF-7 and MCF7-AR1 cells were stored in liquid nitrogen, and new cells were thawed to replace cells that had undergone approximately 25 passages.

#### 3.5.1.2 Proliferation Assay

The methods for the E-screen and A-screen were based on those described by Soto et al. (2004). MCF-7 or MCF7-AR1 cells were seeded in 24-well plates to achieve 20,000-30,000 cells per well. Twenty-four hours after seeding, the media was removed and experimental media (CD-media) was added. The experimental medium was Dulbecco's Modification of Eagles Medium without phenol red (Irvine Scientific, Irvine, Calif.); the 5% FBS used in this medium was stripped of steroids with a charcoal dextran (CD) stripping procedure.

***Standard Curves.*** For the E-screen, each set of experiments was conducted with a 17β-estradiol standard curve consisting of 15 concentrations of 17β-estradiol in CD-media ranging from 0.027 ng/L to 2,724 ng/L. For the A-screen, each set of experiments was conducted with a testosterone standard curve consisting of 15 concentrations of testosterone ranging from 0.29 ng/L to 2,900 ng/L in CD-media supplemented with 17β-estradiol at 27.2 ng/L. The synthetic androgen R1881 is recommended as a standard for the A-screen because unlike testosterone, it is not able to be metabolized (Soto et al., 2004). Because the research compared data with the chemical analyses for testosterone, the researchers chose to use testosterone as the standard. All treatments were done in quadruplicate, with three plates needed to obtain a complete standard curve. Four negative control wells were included on each plate.

***Target Chemicals.*** In order to determine the relative potency of the target chemicals (17β-estradiol, estrone, estriol, and 17α-ethinylestradiol, testosterone, 4-nonylphenol, 4-t-octylphenol, 4-octylphenol, and bisphenol A) using the E-screen, dose-response curves for each of the compounds were made using at least eight different concentrations. Each treatment was done in quadruplicate, and four control wells were included on each plate. Each chemical was tested in at least two independent runs. Because the A-screen assay is run with the cells maximally proliferated with 17β-estradiol, it is not possible to determine if the estrogens have androgen activity.

***Sample Extracts.*** To determine the estrogenic activity of the samples with the E-screen, the sample extract (in ethanol) was added to CD-media at a concentration no greater than 1% and a dilution series was made to ensure active samples fell within the linear portion of the standard curve. Each sample dilution was applied to four wells at a volume of 0.5 mL. 17β-estradiol was added to obtain a concentration of 27.4 ng/L in one of the wells. This positive control was used to determine whether anything in the sample was preventing the growth of the cells. Again, each plate contained four negative control wells.

To determine the androgenic activity of the samples with the A-screen, the sample extract (in ethanol) at a concentration no greater than 1% was added to CD-media supplemented with 17β-estradiol at 27.4 ng/L, and a dilution series was made to ensure active samples fell within the linear portion of the standard curve. Each sample dilution was applied to six wells at a volume of 0.5 mLs. Because cytotoxicity caused by samples and androgenicity will both result in reduced proliferation, in two of the six wells, MCF-7 cells were plated instead of MCF-7AR1 cells. If the sample caused a reduction in proliferation of the MCF-7 cells relative to the controls, the concentration of sample used was considered cytotoxic and was not used to quantify the androgenicity of the sample.

***Proliferation Measurement.*** After five days of incubation, cell proliferation was measured by the sulphorhodamine B (SRB) (Sigma-Aldrich, St. Louis, Mo.) protein assay. The SRB assay determines total cell numbers by measuring total protein content. At the end of the exposure, the media was removed from the cells and 10% trichloroacetic acid (TCA) solution was added to each well to fix the cells. After 20-30 minutes, the TCA was removed, the wells were allowed to dry and the SRB solution (0.4% SRB dye in 1% acetic acid/99% distilled water) was added to each well to stain the cells. The residual SRB dye was removed by rinsing with 1% acetic acid after 20-30 minutes. After drying, the remaining dye was redissolved using 10mM Tris solution (pH 10.5) and read at a wavelength of 515 nm. The plates were read at a wavelength of 515 nm on a Molecular Devices microplate reader (Sunnyvale, Calif.).

### 3.5.1.3 Data Analysis for the E-screen

***Standard Curves***. For the E-screen, the standard curve was fit with a 4-parameter logistic equation with Softmax PRO v. 2.6 software package, the software analysis package for the microplate reader. The response of the MCF-7 cells is sigmoidal; cell numbers are not higher than controls until a concentration of 0.27 ng/L and maximum proliferation occurs at approximately 27 ng/L (Figure 3-5). The assay is highly reproducible, as indicated by the low standard deviations.

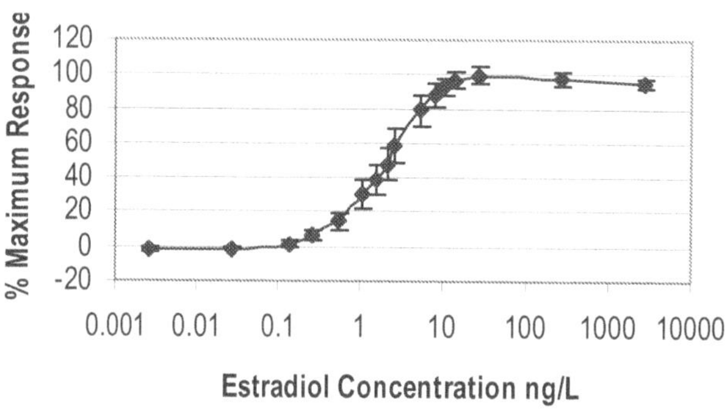


**Figure 3-5. Response of MCF-7 cells to 17β-Estradiol.** Values represent means ± s. d. of 36 separate experiments.

***Target Chemicals.*** The concentration causing 50% of the maximum cell proliferation ($EC_{50}$) for the concentration-response curves of the target compounds was calculated using Microcal Origin v. 4.1 (Northampton, Mass.). The potency for each chemical was determined relative to the $EC_{50}$ of 17β-estradiol. 17β-estradiol and 17α-ethinylestradiol are the most potent chemicals (Table 3-2). The metabolites of 17β-estradiol, estriol, and estrone are 26% and 12% as potent as 17β-estradiol, respectively. The phenolic compounds (nonylphenol, octylphenol and bisphenol A), are five to six orders of magnitude less potent than 17β-estradiol. The researchers were unable to achieve good estimates of the estrogenicity of the phthalates tested because although cells proliferated greater than controls at high concentrations (higher than ~$2 \times 10^7$ ng/L), the maximum proliferation was generally less than 30% of the proliferation caused by 17β-estradiol, and the cells showed signs of toxicity. Additionally, the phthalates were not very soluble in the media at these high concentrations.

Table 3-2. Relative Potencies of Several Target Chemicals for E-screens.

| Chemical | N | $EC_{50}$ ng/L | 95% Confidence interval | Potency Relative to Estradiol |
|---|---|---|---|---|
| 17β-estradiol | 36 | 2.1 | 0.11 | 1 |
| 17α-ethinylestradiol | 3 | 1.8 | 0.17 | 1.2 |
| Estriol | 3 | 8.2 | 0.73 | 0.26 |
| Estrone | 3 | 18 | 1.8 | 0.12 |
| Testosterone | 2 | $1.8 \times 10^5$ | $1.7 \times 10^5$ | $1.1 \times 10^{-5}$ |
| 4-t-Octylphenol | 3 | $2.5 \times 10^4$ | $1.7 \times 10^3$ | $8.5 \times 10^{-5}$ |
| 4-Octylphenol | 3 | $3.4 \times 10^5$ | $2.4 \times 10^5$ | $6.3 \times 10^{-6}$ |
| Nonylphenol | 3 | $4.2 \times 10^4$ | $6.4 \times 10^3$ | $5.1 \times 10^{-5}$ |
| Bisphenol A | 4 | $3.1 \times 10^4$ | $5.6 \times 10^3$ | $6.9 \times 10^{-5}$ |
| di-n-butylphthalate | | $> 2 \times 10^7$ | | |
| di-2-ethylhexylphthlate | | $> 2 \times 10^7$ | | |

***Estradiol Equivalents of Samples.*** To determine the estradiol equivalents (EEq) of the samples, results were interpolated from the standard curve and corrected for the dilution and concentration of the samples. The limit of detection was 0.04 ng/L and was based on 3.29 times the standard deviation of seven blanks (AWWA, 1998). The estrogenic activity expected based on the chemical analyses was determined by multiplying the concentration of the chemical found by the potency factor. The expected EEq for each sample was calculated as the sum of the activities for the analyzed chemicals.

### 3.5.1.4 Data Analysis for the A-screen

***Standard Curves.*** For the A-screen, the data were converted to the percent maximum response relative to the control wells on each plate as follows: [(sample or std $OD_{515}$-minimum $OD_{515}$)/(maximum $OD_{515}$- minimum $OD_{515}$)] × 100, and the testosterone standards were fit with a logistics curve (Microcal Origin). AR1-transfected MCF-7 cells respond to increasing concentrations of testosterone by a decrease in the cell number (Figure 3-6). The reproducibility of the assay was fairly good. The $EC_{50}$ for testosterone averaged 50 ng/L (95% confidence interval = 2.2). Because cells in the A-screen are treated with 17β-estradiol to cause proliferation, it was not meaningful to determine the effects of other estrogens with this assay.

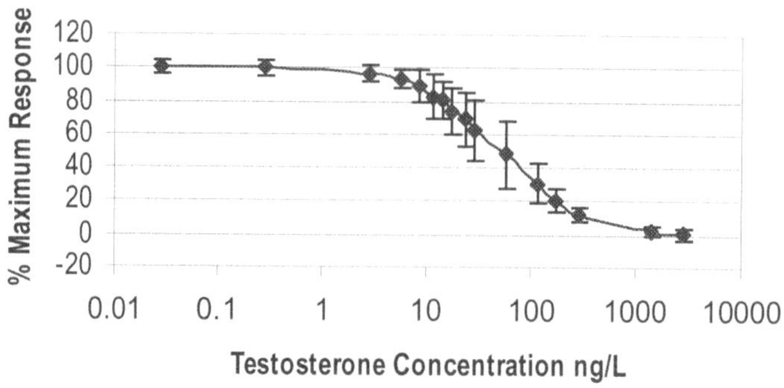


**Figure 3-6. Response of AR1 MCF-7 Cells (A-screen) to Testosterone.**
Values represent means ± s. d. of 15 separate experiments.

***Testosterone Equivalents of Samples.*** Testosterone equivalents (TEq) are interpolated from the standard curve and corrected for the concentration of the sample and dilution used in the assay. Only samples which resulted in less than 80% of the maximum response were considered active. Because of the variability in the curves, the range of the lowest active samples was between one and 5.5 ng/L TEq.

### 3.5.2 Transfected Yeast Assays

#### 3.5.2.1 Description of Transfected Yeast

The transfected yeast cells were obtained from Dr. Gaido at Chemical Industry Institute of Toxicology (Research Triangle Park, N.C.) (Gaido et al., 1997). Two separate yeast strains were used to assess estrogenic and androgenic activity. For estrogenic activity, transformed yeast *Saccharomyces cerevisae* strain BJ3505 contained an expression plasmid with the metallothionein promoter fused to the human estrogen receptor cDNA and a reporter plasmid carrying two estrogen response elements upstream of the β-galactosidase gene. For androgen activity, *S. cerevisae* strain YPH500 strain contained an expression plasmid with the metallothionein promoter fused to the human androgen receptor cDNA and a reporter plasmid carrying two androgen response elements upstream of the β-galactosidase gene and a plasmid encoding for SPT3 to enhance transcriptional efficacy. The assay was performed in a similar matter for both yeast strains.

#### 3.5.2.2 Transfected Yeast Assay

The yeast were grown in selective media to an optical density at 600 nm ($OD_{600}$) ~ 1. The yeast cells were then diluted to an $OD_{600}$ = 0.03 (0.06 for the androgen assay) in selective media supplemented with $CuSO_4$ to induce receptor production and aliquoted into 50-mL conical tubes. The standard curve tubes were dosed with 17β-estradiol at eight concentrations from 0.3-300 ng/L for the ER-yeast assay and with testosterone from 30-30000 ng/L for the AR-yeast assay. Extracts of environmental samples were added at 0.1% to achieve a 1,000-fold dilution of sample in the media. Higher volumes of less active samples were added and evaporated under a gentle stream of $N_2$ to increase sensitivity when necessary but generally were not tested at concentrations greater than a 10-fold dilution. Dilution series were made in the yeast solution, and the tubes were vigorously shaken at 30°C overnight. The following day, the cell concentrations were determined spectrophotometrically, and cells were diluted to a constant yeast concentration ($OD_{600}$ ~ 0.4). A 100-µL aliquot of the yeast mixture was added to a 96-well microtiter plate in triplicate. After the $OD_{590}$ was determined, 100 µL of the *o*-nitrophenyl β-D-galactopyranoside (ONPG) assay mix was added. A yellow color was produced when ONPG was cleaved by the β-galactosidase, and the change in the $OD_{420}$ was determined over 20 minutes.

### 3.5.2.3 Data Analysis

***Standard Curves***. Activity is expressed as $mOD_{420}$/ min divided by cell density ($OD_{600}$). The results are transformed into the % maximum response [(sample or std activity-blank activity)/(maximum activity-blank activity)] × 100, and standards are fit with a logistics curve (Microcal Origin). For the estrogen receptor yeast assay, a standard curve using 17β-estradiol was also run for each batch of samples (Figure 3-7). The response of the yeast is sigmoidal; β-galactosidase activity is not higher than controls until a concentration of about 27 ng/L and maximum activity occurs at approximately 270 ng/L. The response, however, declines at concentrations higher than ~ 270 ng/L. The reproducibility of the assay is fairly good.

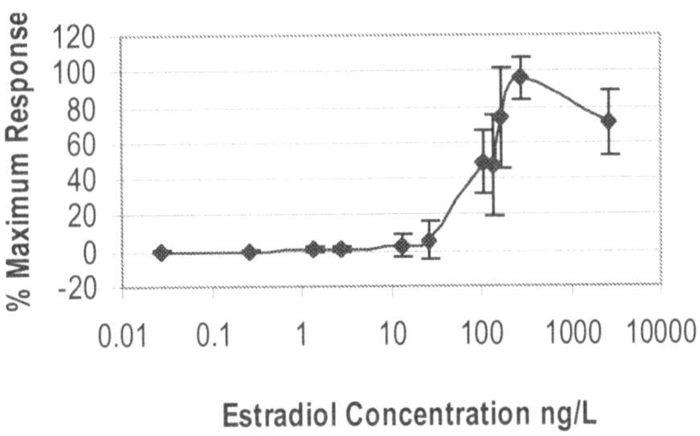


**Figure 3-7. Response of ER-Yeast Cells to 17β-Estradiol.** Values represent means ± s.d. of 18 separate experiments.

The response of the AR-transfected yeast cells to increasing concentrations of testosterone is an increase in β-galactosidase activity (Figure 3-8) and with a similar shape as the response of the ER-yeast cells to 17β-estradiol. The reproducibility of the assay was fairly good, although the sensitivity was markedly lower than for the other bioassays used. The highest $EC_{50}$ among the other bioassays was 120 ng/L (ER-yeast assay), but the $EC_{50}$ for the AR-yeast assay is 4100 ng/L.

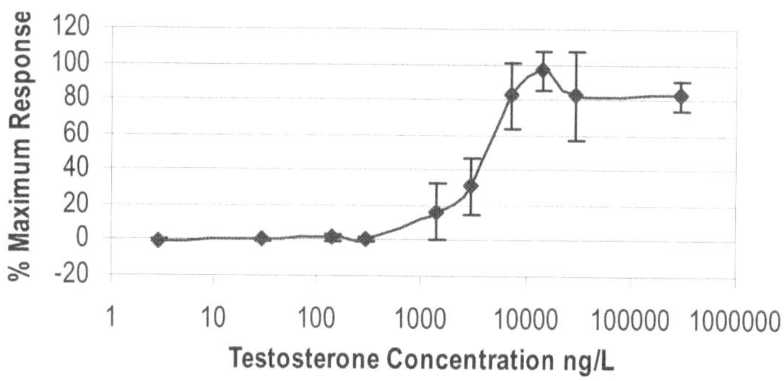


**Figure 3-8. Response of AR-Yeast Cells to Testosterone.** Values represent means ± s. d. 17 separate experiments.

***Target Chemicals***. For the ER-yeast assay, the $EC_{50}$ indicates that 17β-estradiol and 17α-ethinylestradiol are again the most potent chemicals (Table 3-3). Similar to the potencies found with the E-screen, estrone was 11% as potent as 17β-estradiol. In contrast, the relative potency of estriol was much less in the yeast assay (0.75%) than it was for the E-screen (26%). The phenolic compounds (nonylphenol, 4-t-octylphenol, and bisphenol A) are three to five orders of magnitude less potent than 17β-estradiol. Although the relative potency for bisphenol A is similar in both the E-screen and the yeast assay, nonylphenol and 4-t-octylphenol are more than an order of magnitude more potent in the yeast assay than in the E-screen. The phthalates tested did not appear to indicate estrogenicity at the concentrations tested.

**Table 3-3. Relative Potencies of Several Target Chemicals for ER-Yeast Assay.**

| Chemical | N | $EC_{50}$ ng/L | 95% Confidence Interval | Potency Relative to Estradiol |
|---|---|---|---|---|
| 17β-estradiol | 18 | 120 | 7.0 | 1 |
| 17α-ethinylestradiol | 3 | 97 | 4.9 | 1.3 |
| Estriol | 4 | $1.6 \times 10^{4}$ | 690 | 0.0075 |
| Estrone | 3 | 1100 | 620 | 0.11 |
| Testosterone | | Not determined | | |
| 4-t-Octylphenol | 3 | $5.1 \times 10^{4}$ | 5200 | $2.4 \times 10^{-3}$ |
| 4-Octylphenol | 2 | $> 1 \times 10^{6}$ | | |
| Nonylphenol | 5 | $2.0 \times 10^{5}$ | 8600 | $6.0 \times 10^{-4}$ |
| Bisphenol A | 3 | $3.5 \times 10^{6}$ | $1.2 \times 10^{6}$ | $3.5 \times 10^{-5}$ |
| di-n-butylphthalate | | $> 3 \times 10^{8}$ | | |
| di-2-ethylhexylphthlate | | $> 3 \times 10^{8}$ | | |

Estrogens tested in the AR-yeast assay did exhibit activity in this assay; however, the relative potency was much lower than for testosterone and therefore unlikely to account for a substantial proportion of activity seen in actual samples (Table 3-4).

**Table 3-4. Relative Potencies of Several Hormones for AR-Yeast Assay.**

| Chemical | N | $EC_{50}$ ng/L | 95% Confidence Interval | Potency Relative to Testosterone |
|---|---|---|---|---|
| Testosterone | 17 | 4100 | 470 | 1 |
| 17β-estradiol | 4 | $1.3 \times 10^5$ | $1.9 \times 10^4$ | 0.031 |
| 17α-ethinylestradiol | 3 | $5.8 \times 10^6$ | $1.1 \times 10^6$ | $7.1 \times 10^{-4}$ |
| Estriol | 3 | $>2.9 \times 10^7$ | | |
| Estrone | 4 | $9.3 \times 10^6$ | $1.5 \times 10^6$ | $4.4 \times 10^{-4}$ |

***Estrogen and Testosterone Equivalents of Samples.*** EEq and TEq were determined by comparison with the appropriate standard curve and corrected for the concentration of the sample and dilution used in the assay. Samples were considered active if the β-galactosidase activity was higher than the lowest standard with activity higher than the controls. For the ER-yeast assay, the lowest samples considered active ranged from 0.4-5 ng/L (the large range is partially due to some samples being run at a higher concentration). The lowest samples considered active for the AR-yeast assay ranged from 90-218 ng/L.

# CHAPTER 4.0

# METHOD COMPARISON

## 4.1 Objectives of Method Development

To establish the analytical techniques as appropriate and effective, the recoveries of spiked matrices must be determined to establish adequate extraction techniques and give an indication of the precision of the analytical techniques. In addition, the level of contamination must be established by analyzing blanks. Finally, comparing the results of the analytical techniques helps provide insight into strengths and weaknesses of each technique.

## 4.2 Methods

### 4.2.1 Sampling and Analytical Methods

Samples were collected and processed as described in Chapters 2.0 and 3.0.

### 4.2.2 Spike Recovery Study

A study was conducted to compare spike recoveries among bioassays, GC-NCI-MS, and ELISA using a relevant matrix. A 10-mg/L dissolved organic carbon solution of effluent organic matter (EfOM) was spiked with either a low concentration standards mix or a high concentration standards mix (Table 4-1). Nine liters of each sample (EfOM alone, EfOM + low spike, EfOM + high spike) were produced and samples were split between WSLH and CSM. At each location, three separate extractions were completed for each treatment, and analyses were conducted as previously described.

**Table 4-1. Concentrations of Target Chemicals Initially Present in the EfOM and Added to EfOM in Spiking Experiments.**

| Standards | EfOM (ng/L)[1] | Low (ng/L) | High (ng/L) |
|---|---|---|---|
| 17α-ethinylestradiol | 2.5 | 7.9 | 62.9 |
| 17β-estradiol | 13.0 | 7.3 | 58.6 |
| Estriol | < 1 | 8.6 | 68.9 |
| Estrone | < 1 | 9.8 | 78.5 |
| Testosterone | < 1 | 7.6 | 60.5 |
| Nonylphenol | 4500 | 2040 | 20400 |
| 4-t-Octylphenol | 395 | 418 | 4180 |
| 4-Octylphenol | <12 | 125 | 1250 |
| Bisphenol A | 8.3 | 146 | 1460 |

[1]Initial EfOM concentrations as measured by GC-NCI-MS.

## 4.3 Results and Discussion

### 4.3.1 Quality Control/Quality Assurance

#### 4.3.1.1 Blanks

Results from the GC-NCI-MS analyses indicated that the phenolic compounds were present in the Type I water sample extracts. The concentration (means ± standard deviation in ng/L) for the blanks from the majority of sampling events for the phenols were 59 ± 53 for nonylphenol, 24 ± 25 for 4-t-octylphenol, 4 ± 6 for 4-octylphenol, and 39 ± 47 for bisphenol A. Results presented for facility samples are after subtraction of the average blank levels. The hormones were below the limit of detection except for blanks from two facilities. For those two facilities, results were determined after subtracting the blank levels for the corresponding sampling event.

With one exception, the blanks were not active in the bioassays. One extract from a Type I water blank was slightly active in the A-screen.

#### 4.3.1.2 Spike Recoveries

The recoveries for hormones spiked into EfOM as analyzed by GC-NCI MS averaged between 74% to 117% for the high spike samples and 75% to 143% for the low spike samples (Figure 4-1). The replicate analyses indicate that the precision is not as good as would have been desired for the low hormone concentration but fairly robust for the high hormone spikes. With one exception, the recoveries for the phenols were quite good, with reasonable precision for all compounds (Figure 4-2). The less than ideal recoveries for the low spike of nonylphenol is likely because of the high concentration of nonylphenol (~ 4500 ng/L) found in the EfOM.

The comparison between the ELISA and the GC-NCI-MS results indicate that the recoveries are generally higher for the GC-NCI-MS than for the ELISA analyses (Figure 4-3). With the exception of the GC-NCI-MS results for the low spike of estradiol, the standard deviations indicate that the precision is reasonably good.

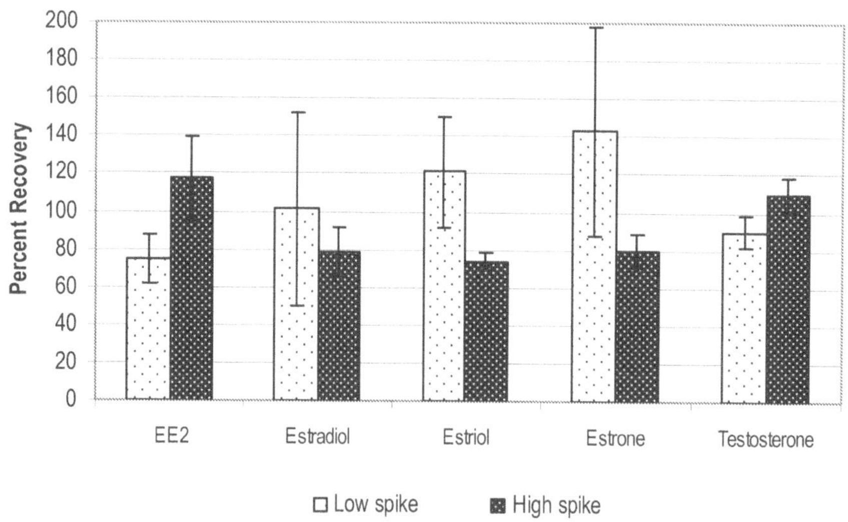


**Figure 4-1. GC-NCI-MS Recoveries of Hormones of the EfOM Spike Study.**
Values represent the means ± s.d. for three separate extractions. EE2 = 17α-ethinylestradiol.

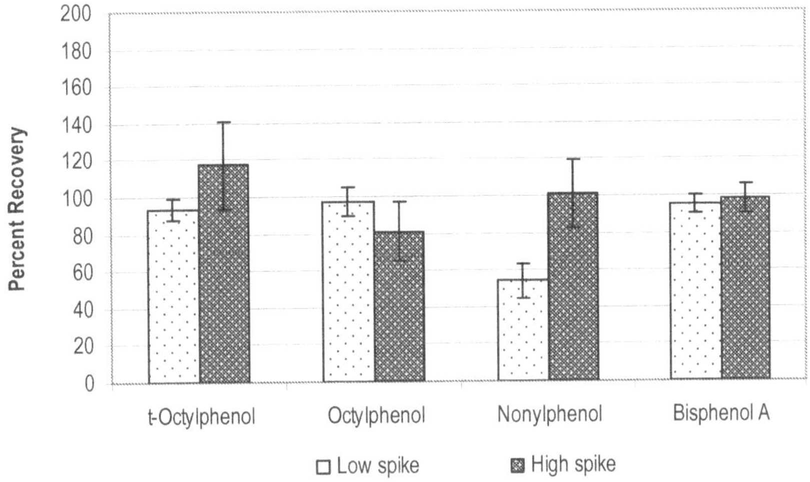


**Figure 4-2. Comparison Between GC-NCI-MS Recoveries of Phenols for the EfOM Spike Study.**
Values represent the means ± s. d. for three separate extractions.

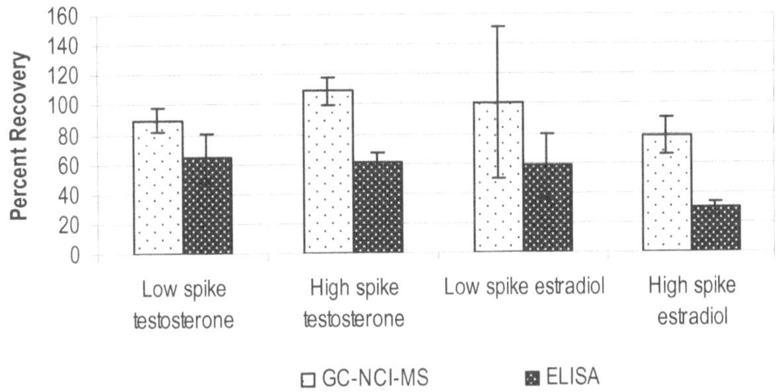


**Figure 4-3. Comparison Between GC-NCI-MS and ELISA Recoveries for the EfOM Spike Study.**
Values represent the means ± s. d. for three separate extractions.

Results from three of the four bioassays indicate good agreement with predicted activities, with recoveries from 80% to 127% of the expected activity (Figure 4-4). The AR-yeast assay results indicated lower recoveries, with the low spike samples being only 42%. The replicates seemed to be in good agreement, with the standard deviations less than 15% of the mean in all assays.

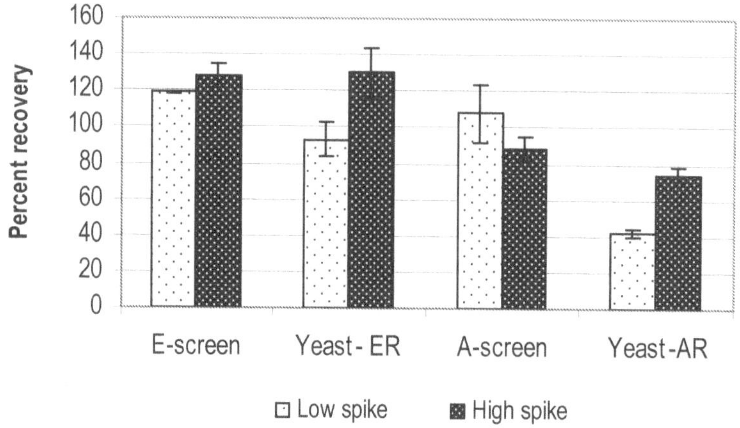


**Figure 4-4 Recoveries for Bioassays for the EfOM Spike Study.**
Recoveries are based on estradiol equivalents of the spiking mixtures for the E-screen and ER-yeast assay and testosterone equivalents for the A-screen and AR-yeast assay. Values represent the means ± s. d. for three separate extractions.

#### 4.3.1.3 Duplicates

A duplicate extraction was planned for each sampling event; however, bottle breakage and other unforeseen problems meant that duplicates could not be completed for each sampling event. Overall, eight facility samples were extracted in duplicate. Average relative percent differences ranged from 5% to 172% for the GC-NCI-MS samples, 5% to 48% for ELISA analyses, and 1% to 30% for bioassay duplicates (Table 4-2).

Table 4-2. Relative Percent Differences for Duplicate Extractions.

| Analytical Technique | n | Relative Percent Difference: Range | Relative Percent Difference: Average |
|---|---|---|---|
| **GC-NCI-MS** | | | |
| 17β-estradiol | 2 | 31 - 41 | 38 |
| 17α-ethinylestradiol | 1 | 71 | 71 |
| Estriol | 3 | 11 - 23 | 16 |
| Estrone | 6 | 11 - 170 | 54 |
| Testosterone | 4 | 15 – 94 | 62 |
| 4-t-Octylphenol | 6 | 8 – 66 | 35 |
| 4-Octylphenol | 2 | 6 – 172 | 89 |
| Nonylphenol | 6 | 5 - 110 | 42 |
| Bisphenol A | 5 | 10 - 88 | 34 |
| **ELISA** | | | |
| 17β-estradiol | 2 | 5 – 48 | 27 |
| Estriol | 2 | 5 – 26 | 16 |
| Testosterone | 2 | 26 – 40 | 33 |
| **Bioassays** | | | |
| E-screen | 5 | 1 – 23 | 12 |
| ER Yeast | 4 | 2 – 30 | 13 |
| A-screen | 3 | 15 – 23 | 18 |
| AR Yeast | 2 | 11 – 19 | 15 |

Note: Relative percent differences were not determined for samples below detection limits and therefore were not included in duplicate analyses.

## 4.3.2 Comparison of Results Among Analytical Techniques

### 4.3.2.1 Comparing GC-NCI-MS to ELISA

Overall, the relationships between results obtained with the two chemical analytical techniques were inconsistent (Figure 4-5). The results for 17β-estradiol and estrone were within the same order of magnitude for the two techniques, and many samples indicated similar results, especially for samples near or below detection limits. With respect to estriol, GC-NCI-MS results were an order of magnitude higher than were the results from ELISA. The reasons for the discrepancies are not clear. One possibility is that there are many differences in the extraction methods, although both methods were thought to optimize analytical results. The researchers attempted to split some extracts generated with ELISA, but the GC-NCI-MS did not find any hormones in the extracts tested. Recoveries in the spike recovery study indicate that ELISA is more technically challenging. Organic matter interference can result in false positives in an ELISA (Huang and Sedlak, 2001). Conversely, incomplete collection of the peak of interest will result in underestimates.

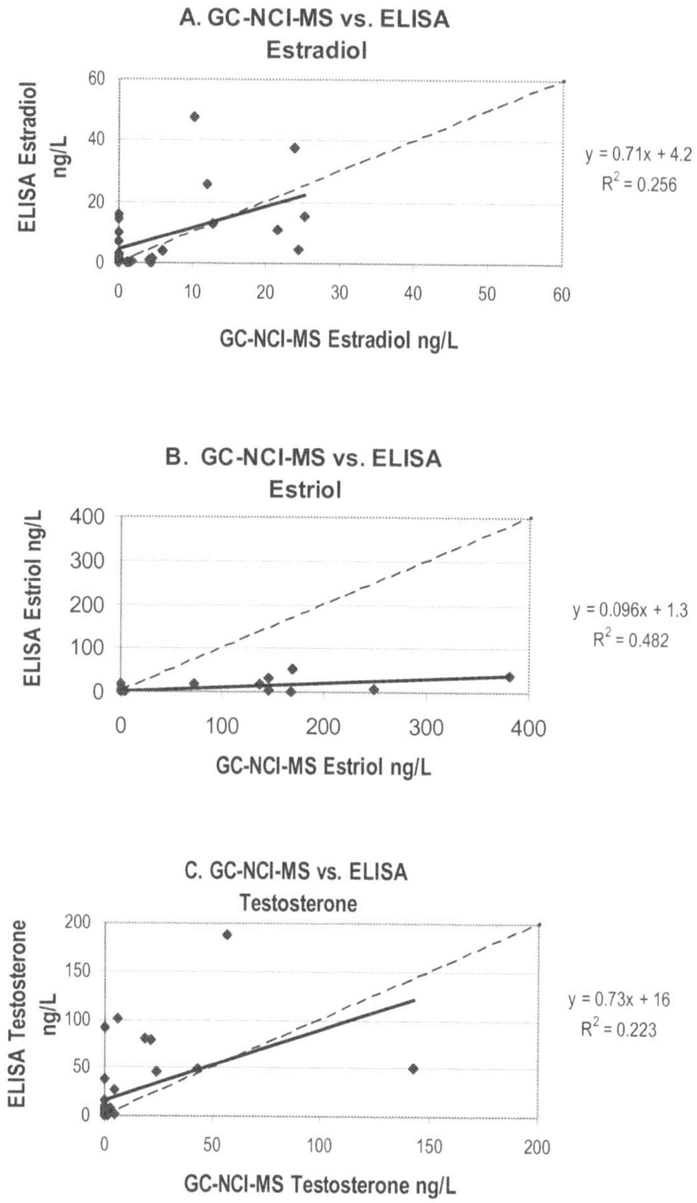


**Figure 4-5. Comparison Between GC-NCI-MS and ELISA for A. 17β-Estradiol, B. Estriol, and C. Testosterone.** The dashed line represents a 1:1 relationship. The solid line was generated with regression analysis comparing the two data sets and is represented by the listed equation.

### 4.3.2.2 Comparing GC-NCI-MS to Estrogenic Bioassays

Overall, the comparison between the GC-NCI-MS results converted to 17β-estradiol equivalents and the estrogenic bioassays (the E-screen and ER-yeast) indicated good agreement between the methods (Figure 4-6). The relationship was slightly stronger with the E-screen ($r^2 = 0.86$) than with the ER-yeast ($r^2 = 0.76$). With the E-screen, the slope is reasonably close to 1, indicating little bias between the methods. The slope of the ER-yeast with the GC-NCI-MS results is somewhat lower than 1, which might be due to organic matter interferences in the yeast assay (see Appendix F).

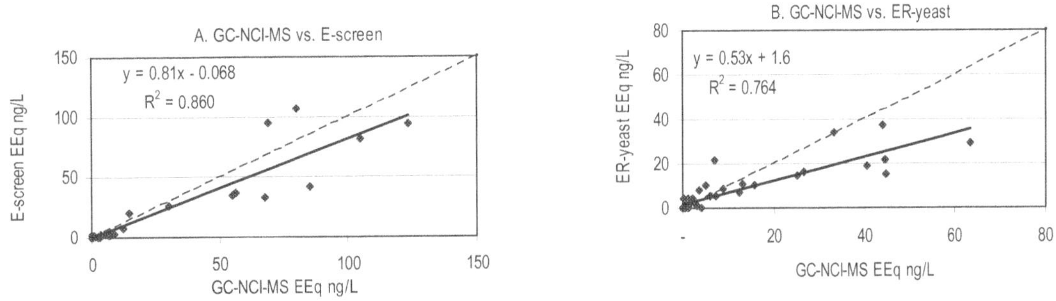


**Figure 4-6. Comparison Between GC-NCI-MS and A. E-screen and B. ER-Yeast Assay.**
The dashed line represents a 1:1 relationship. The solid line was generated with regression analysis comparing the two data sets and is represented by the listed equation.

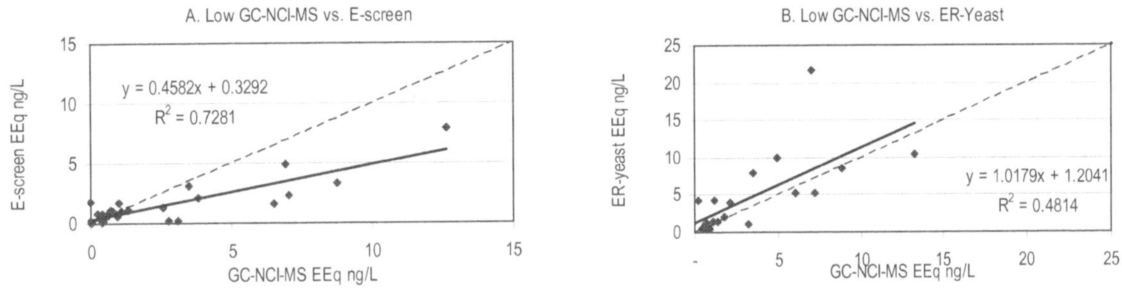


**Figure 4-7. Comparison Between GC-NCI-MS and A. E-screen and B. ER-Yeast Assay on the Treated Wastewater Samples.** The dashed line represents a 1:1 relationship. The solid line was generated with regression analysis comparing the two data sets and is represented by the listed equation.

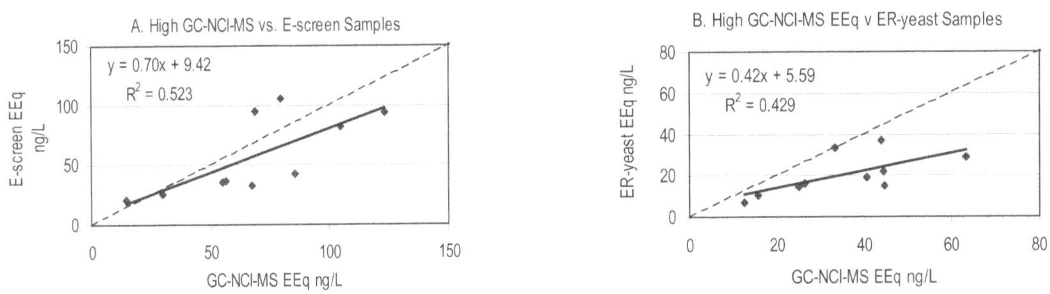


**Figure 4-8. Comparison Between GC-NCI-MS and A. E-screen and B. ER-Yeast Assay on the Least Treated Samples.** The dashed line represents a 1:1 relationship. The solid line was generated with regression analysis comparing the two data sets and is represented by the listed equation.

The data sets from the cleanest (low) samples (Figure 4-7) and the least treated (high) samples (Figure 4-8) were examined individually to determine whether the comparison was adequate over both ranges of sample complexity. There was good agreement between the GC-NCI-MS and E-screen for the low samples ($R^2 = 0.73$), although the GC-NCI-MS predicts higher values than were seen in the bioassay (slope = 0.46). The agreement between the GC-NCI-MS and the ER-yeast assay was not as strong ($R^2 = 0.48$); however, the slope was very near 1, showing similar results for both analyses. One point (ER-yeast = 21 ng/L EEq) may be an outlier, which if removed would improve the $R^2$ to 0.65, but the slope would be lowered to 0.78. Looking at the least treated (high) samples, the variability is slightly higher for both assays, although there is closer agreement between the results for the E-screen, with the slope closer to 1 (0.70). The results for the yeast assay are lower than predicted by the chemical analyses.

#### 4.3.2.3 Comparing ELISA with Estrogenic Bioassays

The relationships between the estrogen bioassays and predicted activity based on the ELISA EEq were poor (Figure 4-9). Although only two compounds were measured with the ELISA (17β-estradiol, estriol), those two make up a substantial proportion of the activity for most of the samples in the E-screen and therefore would be expected to relate better than $R^2 = 0.36$. The yeast assay would not be expected to relate well to the ELISA results because estriol is not potent in the ER-yeast assay. This is consistent with the low $R^2$ that was found (0.14).

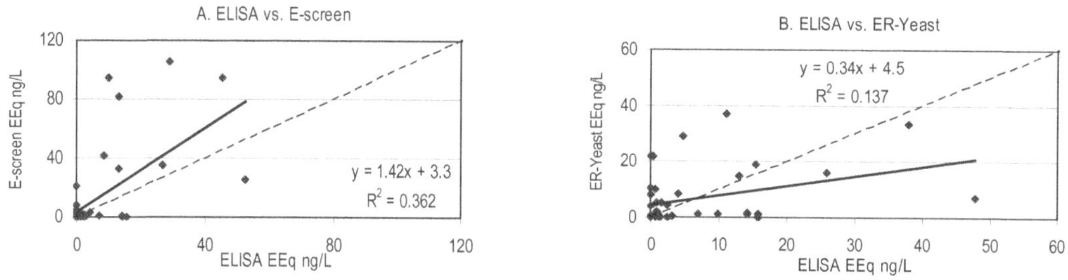


**Figure 4-9. Comparison Between ELISA and A. E-screen and B. ER-Yeast Assay.**
The dashed line represents a 1:1 relationship. The solid line was generated with regression analysis comparing the two data sets and is represented by the listed equation.

#### 4.3.2.4 Comparing the E-screen to the ER-Yeast Assay

The E-screen and ER-yeast assay are not expected to give identical results because of the difference in potency of chemicals in the two different assays (Figure 4-10). Thus, based on chemical analyses, the yeast assay would be expected to be approximately half as active as the E-screen. Overall, the relationship is quite strong ($R^2 = 0.7$). The observed data, however, indicated somewhat lower overall activity in the ER-yeast assay, as the slope was 0.29, which is less than the 0.47 predicted.

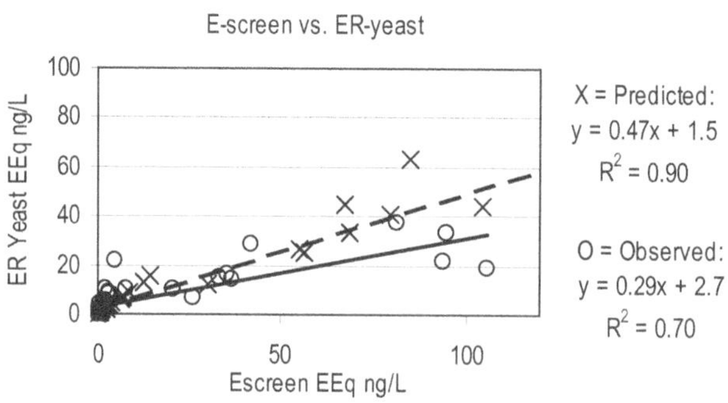


**Figure 4-10. The Relationship Between the E-screen and ER-Yeast Bioassays.**
The predicted values (x) are the EEq calculated by the GC-NCI-MS results. The dashed line indicates the regression of the predicted data set. The solid line represents the regression of the observed values (o).

### 4.3.2.5 Comparing Androgen Analyses

The relationships between testosterone concentrations and androgen activity for all influent and effluent samples combined were not robust (Figure 4-11). The ELISA results exhibited a slightly tighter relationship with the bioassays than did the GC-NCI-MS results. The bioassays indicated that higher testosterone concentrations were expected than actually indicated based on either chemical analytical technique. In this case, it is very likely that androgens other than testosterone wee present that contributed to the activity.

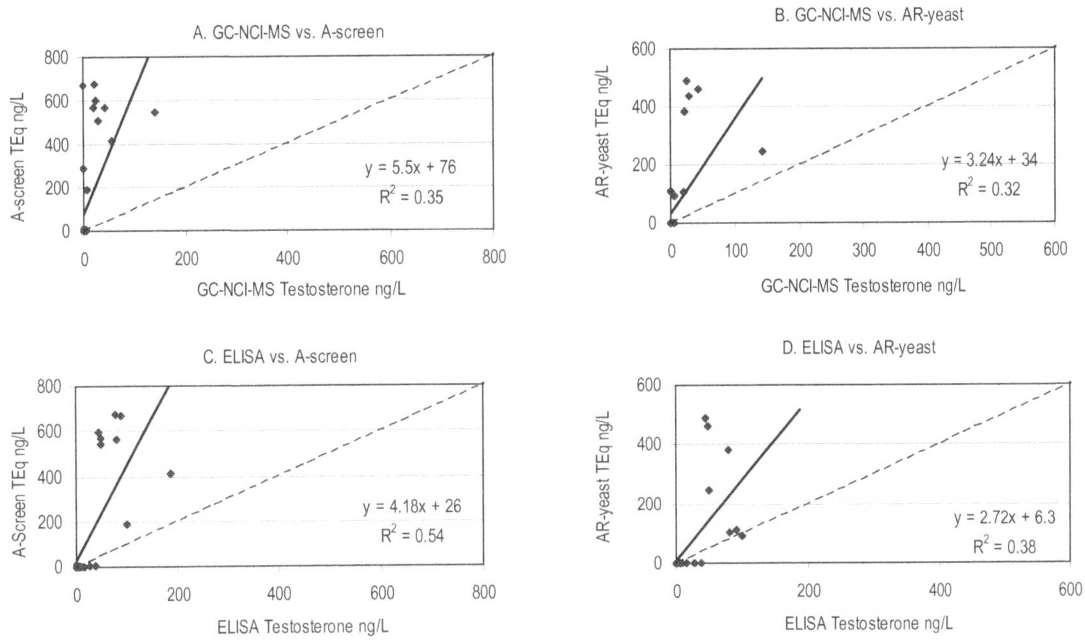


**Figure 4-11. Relationships Between Testosterone Analyses and Androgen Bioassay Activities.**
The dashed line represents a 1:1 relationship. The solid line was generated with regression analysis comparing the two data sets and is represented by the listed equation. A. compares GC-NCI-MS testosterone results with the A-screen; B. Compares GC-NCI-MS testosterone results with the AR-yeast. C. compares ELISA testosterone results with the A-screen; D. Compares ELISA testosterone results with the AR-yeast.

The relationship between the results of the A-screen and the AR-yeast assay is not strong (Figure 4-12). The comparison between the A-screen assay and the AR-yeast assay is weakened by the high detection limits of the AR-yeast assay. Thus only minimally treated samples exhibited any activity. All the values for the yeast assay fell below the 1:1 line expected. As all of the chemical components have not been characterized between the two assays or quantified chemically, the research team can not know how much of this is due to difference in potencies, or how much to the inhibition by organic matter (see Appendix F). Two of the samples were below detection for the AR-yeast assay but, according to the A-screen, should have been high enough to have positive results for the yeast assay. However, the results are generally in the same order of magnitude and are worthy of continued exploration.

The relationships between testosterone concentrations and androgen activity were not robust (Figure 4-11). The ELISA results exhibited a slightly tighter relationship with the bioassays than did the GC-NCI-MS results. The bioassays indicated that higher testosterone concentrations were expected than actually indicated based on either analytical technique. In this case, it is very likely that androgens other than testosterone were present that contributed to the activity.

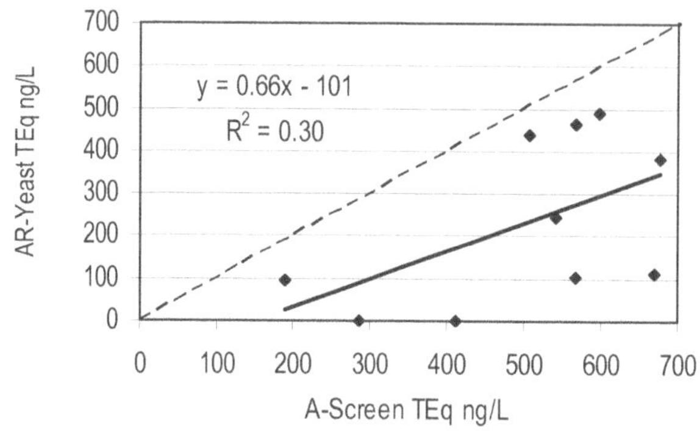


**Figure 4-12. Relationship Between A-screen and AR-Yeast.**
The dashed line represents a 1:1 relationship. The solid line was generated with regression analysis comparing the two data sets and is represented by the listed equation.

## 4.4 Conclusions

Working with the complex matrices in wastewater offers many analytical challenges. QA samples indicated blank contamination, especially with phenolic compounds. Recoveries were reasonable, although not spectacular, with variability often quite high for the chemical analyses. The ranges found for the duplicates were also higher than would be desirable but good enough to indicate reasonable estimates of the results. Given the issues described and variability seen, the strong relationships found for the estrogenic bioassays and GC-NCI-MS indicates that all three analytical techniques are likely acceptable approaches with which to study the removal of estrogenic compounds and activity during wastewater treatment. Although ELISA analyses had initially been expected to be a simple and inexpensive analytical technique, the cleanup needed to remove organic matter made the technique more cumbersome and apparently less robust than was anticipated.

# CHAPTER 5.0

# CHANGES IN EDCS AND ACTIVITY AFTER TREATMENT WITH ACTIVATED SLUDGE

## 5.1 Introduction

An increasing number of utilities are recycling treated wastewater for beneficial reuse in nonpotable and potable applications. Results from previous research indicate that municipal wastewater effluents may contain compounds capable of endocrine disruption (Purdom et al., 1994; Harries et al., 1996; Folmar et al., 1996; Rodgers-Gray et al., 2000). Endocrine disrupting compounds (EDCs) can be divided into three major classes according to their endpoints: 1) estrogenic compounds that mimic or block natural estrogens, 2) androgenic compounds that mimic or block natural testosterone, and 3) thyroidal compounds that cause direct or indirect impacts on the thyroidal glands.

More than 70,000 chemicals are believed to have endocrine disruptive potential (Gillesby and Zacharewski, 1998). Hormonally active agents (17β-estradiol, estrone, estriol, testosterone, and the synthetic hormone 17α-ethinylestradiol in contraceptives) exhibit a high endocrine disrupting potency for various endpoints (Snyder et al., 2001). Some other chemical groups that can cause endocrine disruption and have been reported to occur in sewage effluent include alkylphenolpolyethoxylates (APECs) and their degradation products, bisphenol A, and some phthalates (Jobling and Sumpter, 1993; Montgomery-Brown et al., 2003). The endocrine disrupting potency of the above compounds is appreciably lower as compared to steroid hormones (Fang et al., 2000).

Nevertheless, because of their prevalence in domestic sewage, EDCs might survive treatment during water reclamation, resulting in residual concentrations in receiving waters that have the potential to cause adverse effects on human or aquatic life (Hemming et al., 2003). Steroidal hormones, especially 17α-ethinylestradiol and 17β-estradiol, are often present in wastewater effluents at concentrations above levels shown to cause adverse endocrine effects in fish (Belfroid et al., 1999; Baronti et al., 2000). However, because of the difficulty of measuring trace chemicals in sewage effluent, data are often viewed skeptically since different analytical methods were employed, and method detection limits can vary by an order of magnitude (Snyder et al., 1999; Ternes et al., 1999a; Huang and Sedlak, 2001). Further, mechanisms responsible for the attenuation of compounds causing endocrine disruption during wastewater treatment or water reclamation are not well understood. Conventional wastewater treatment facilities are not specifically designed to remove EDCs, and the degree to which they are removed varies from nearly complete to very little (Ternes et al., 1999a; Baronti et al., 2000; Holbrook et al., 2002). There is a lack of understanding whether observed variations of target compound concentrations and activities are a result of the efficiency of a single unit process, operational parameters (such as organic carbon loading, hydraulic (HRT) and solid retention times (SRT)), or sequences of unit processes employed during water reclamation.

The results of multiple studies show that adsorption onto suspended solids, aerobic and anaerobic biodegradation, chemical (abiotic) degradation via processes such as hydrolysis, and volatilization are the primary removal mechanisms for EDCs in wastewater treatment processes (Holbrook et al., 2002; Birkett and Lester 2003, Mansell et al., 2004). Photolytic decay has also been suggested as a removal mechanism for steroid hormones during exposure times exceeding 24 hours (Gray and Sedlak, 2003; Mansell et al., 2004). Many studies have suggested that increased SRTs improve the removal of EDCs in wastewater treatment plants (Holbrook et al., 2001; Andersen et al., 2003). Hydraulic retention time has also been reported to have a positive correlation with EDC removal (Strenn et al., 2003). Finally, there is a need to study removal, modification, or inactivation of EDCs and their metabolites using multiple techniques (including bioassays) to confirm chemical measurements and to assure that chemical concentrations are linked to biological responses. Given the above, the scope of this study was to combine bioassays and chemical measurements to assess the efficacy of water reclamation processes (with special focus on secondary treatment) in removing estrogenic disrupting activity and androgens (testosterone) from treated effluents. During the overall study, five different bioassays were employed: ER-yeast, AR-yeast, human breast cell cancer assays for estrogenic (E-screen) and androgenic endpoints (A-screen), and a tadpole metamorphosis assay for a thyroidal endpoint. Results presented in this chapter do not include a thyroid endpoint.

## 5.2 Methods

The target chemicals, extraction methods and analytical techniques were conducted as described in Chapter 3.0.

Field investigations reported here focused on seven full-scale water reclamation facilities located in California, Arizona, Florida, Virginia, and Wisconsin and are described fully in Chapter 2.0. Using these facilities permitted an investigation of both regional variations of water quality and the removal effectiveness of individual unit operations comprising an overall process train. Prior to sampling, staff of each facility provided detailed information on the service area characteristics, treatment process, operational parameters, and water quality data for the past four years. Technical details and operational parameters of each facility are summarized in Table 5-1. Flow schematics of the secondary treatment process for each full-scale facility are illustrated in Appendix A (Figures A-1 through A-12). All full-scale facilities utilize activated sludge (AS) treatment; some employ partial nitrification and nitrification/denitrification as well as chemical or biological phosphorus removal. HRTs of the full-scale secondary treatment processes varied between 1.2 to 13.7 hours, with SRTs ranging from 1.7 to 10 days.

Table 5-1. Operational Parameters for Activated Sludge Facilities.

| Facility | Wastewater Treatment | Design Flow ($m^3/d$) | Sampling Flow ($m^3/d$) | HRT (hours) | SRT (days) | MLSS (mg/L) | Organics Loading BOD[1] (kg/day) | Influent Effluent BOD[1] (mg/L) | Influent Effluent $NH_3$-N[1] (mg/L) | Effluent $NO_3$-N[1] (mg/L) |
|---|---|---|---|---|---|---|---|---|---|---|
| A | Nitrification/Bio-P removal | 94,636 | 76,900 | 13.7 | 10.0 | 1,900 | 10,689 | 139<br>4.1 | 23.5<br>0.02 | 13.1 |
| C | Pre-aeration/activated sludge/chlorination | 136,275 | 104,894 | 1.2 | 1.7 | 2,180 | 18,042 | 172<br>4 | 30<br>33 | <0.2 |
| D | Nitrification/denitrification/ tertiary filtration/UV-disinfection | 236,588 | 237,270 | 2.3 | 10.0 | 3,660 | 61,690 | 260<br>2 | 34.5<br>2.4 | 6.64 |
| E | Nitrification/denitrification/ tertiary filtration/UV-disinfection | 68,137 | 34,447 | 5.0 | 10.0 | 3,000 | 6407 | 186<br>3 | 25.5<br>0.13 | 6.59 |
| F | Partial nitrification/tertiary filtration/chlorination | 87,064 | 68,932 | 6.8 | 4.4 | 2,229 | 9,857 | 143<br>4 | 28.5<br>4.74 | 2.05 |
| G | Nitrification/denitrification/ tertiary filtration/ chlorination ("CA Title 22")/dechlorination | 238,481 | 238,481 | 7.0 | 7.0 | 2,000 | 41,019 | 172<br>2.2 | NA<br>0.46 | 5.26 |
| H1 | Nitrification/denitrification/ tertiary filtration/UV-disinfection | 632,164 | 480,747 | 4.0 | 7.0 | 3,000 | 67,305 | 140<br>8 | 24.8<br><0.1 | 6.1 |
| H2 | Nitrification/denitrification/ tertiary filtration/UV-disinfection | 632,164 | 480,747 | 5.0 | 9.0 | 3,000 | 67,305 | 140<br>5 | 24.8<br><0.1 | 6.5 |

NA – not available MLSS – Mixed liquor suspended solids
[1]Water quality observed during sampling campaign.

## 5.3 Results

All facilities investigated in this study received untreated domestic wastewater from urban settings located in different areas of the United States. The design capacity of the full-scale facilities ranged from 68,000 to 238,500 $m^3$/day. The untreated sewage at all sites was characterized as medium strength wastewater based on $BOD_5$, ammonia nitrogen, and phosphorus concentrations (Metcalf and Eddy, 2003). During the sampling campaigns, all facilities achieved a BOD removal exceeding 94% and nitrogen removal similar to values determined by the initial utility survey indicating performance within design specifications (Table 5-1). Sampling of secondary treated effluents occurred prior to disinfection.

### 5.3.1 Occurrence of Estrogenic EDCs in Primary and Secondary Effluents

The concentrations of estrogenic steroid target compounds observed in the primary and secondary effluents are summarized in Table 5-2. Concentrations of individual target compounds were quantified using GC-NCI-MS and HPLC-ELISA. 17β-estradiol and testosterone concentrations, which were determined with both analytical methods, showed good agreement for some of the secondary effluent samples and for most of the primary effluent samples. Primary effluent samples for estriol, however, exhibited significant differences between the two methods, which might be attributed to organic matter matrix effects while processing primary treated wastewater samples. While both analytical methods had similar detection limits for 17β-estradiol (<1.3 and < 0.6 ng/L for GC-NCI-MS and ELISA, respectively), the HPLC-ELISA method for estriol and testosterone indicated better sensitivity at the low concentration range with a detection limit below 1 ng/L for both compounds.

17α-ethinylestradiol was only detected in one out of seven primary effluents (Facility C) at a concentration of 14.4 ng/L. This facility was also the only one with a concentration in the secondary effluent exceeding the detection limit of 0.7 ng/L. Facility C exhibited the shortest HRT and SRT of all facilities investigated. 17β-estradiol was not observed in the primary effluent for facility A but was found in all other primary effluents above 12 ng/L. Estriol and estrone were prevalent in every primary effluent in concentrations exceeding 32 ng/L and 8 ng/L, respectively, based on GC-NCI-MS measurements. Estrone, which is readily formed by oxidation of 17β-estradiol in activated sludge (Weber et al., 2005), exhibited the highest concentration of all steroid hormones in secondary treated effluents.

The total estrogenic activity quantified through the E-screen varied in primary effluents from 20 to 95 ng/L EEq. For nitrifying/denitrifying facilities, the 17β-estradiol equivalent concentration in secondary treated effluents ranged from 0.18 to 4.9 ng/L. The highest E-screen concentration, 7.9 ng/L, was determined in the secondary effluent of Facility C, which exhibited the shortest SRT and did not nitrify. The estrogenic activity quantified through the ER-yeast assay indicated a lower range of activity in the primary effluent relative to the E-screen, between 7 and 33 ng/L EEq. However, the secondary effluents assayed with the ER-yeast indicated a wide range in the samples, from <0.4 to 21.6. The 21.6 may be an outlier, as it is not in the same range as the other results. The second highest was 10.5 ng/L from Facility C. In general, the occurrence of EDCs did not show a regional pattern considering the different locations and climates of the various facilities.

Table 5-2. Occurrence of Estrogenic Activity and Target Hormones in Primary and Secondary Treated Effluents in ng/L.

| | | 17β-EEq | 17β-EEq | 17α-ethinyl-estradiol | 17β-estradiol | | Estriol | | Estrone |
|---|---|---|---|---|---|---|---|---|---|
| Facility | Method | E-Screen | ER-Yeast | GC-NCI-MS | GC-NCI-MS | HPLC-ELISA | GC-NCI-MS | HPLC-ELISA | GC-NCI-MS |
| A | Prim. effluent | 20.8 | 10.3 | <0.7 | <1 | NA | 32.7 | NA | 46.3 |
| | Sec. Effluent | 0.82 | 0.44 | <0.7 | <1 | NA | <3.3 | NA | 2.0 |
| C | Prim. effluent | 42 | 28.8 | 14.4 | 24.5 | 4.6 | 138 | 15.8 | 60 |
| | Sec. effluent | 7.91 | 10.5 | 4.1 | 4.4 | <0.6 | 4.0 | <0.4 | 17.7 |
| D | Prim. effluent | 94.7 | 33.6 | <0.7 | 23.9 | 37.6 | 147 | 29.9 | 56.0 |
| | Sec. effluent | 4.93 | 21.6 | <0.7 | 1.5 | <0.6 | <4.7 | <0.4 | 46.0 |
| E | Prim. effluent | 94.1 | 21.6 | <0.7 | 15.3 | NA | 381 | 38.3 | 65.2 |
| | Sec. effluent | 3.26 | 8.6 | <0.7 | 6 | 4.1 | 4.9 | 0.7 | 11.1 |
| F | Prim. effluent | 35.1 | 16.2 | <0.7 | 12 | 25.9 | 146 | 2.7 | 37.4 |
| | Sec. effluent | 0.18 | < 0.5 | <0.7 | <0.6 | 0.69 | <3.3 | <0.4 | 25.6 |
| G | Prim. effluent | 32.9 | 15.1 | <0.7 | 12.8 | 13.1 | 168 | <0.4 | 80.3 |
| | Sec. effluent | 2.12 | 10 | <0.7 | 1.8 | 0.62 | <3.3 | <0.4 | 16.4 |
| H | Prim. effluent | 25.9 | 7.06 | <0.7 | 14.1 | 47.8 | 72 | 18.9 | 8.4 |
| | Sec. effluent 1 | 1.12 | 1.47 | <0.7 | <0.6 | 14.2 | <3.4 | <0.4 | 5.6 |
| | Sec. effluent 2 | 1.07 | 2.02 | <0.7 | <0.6 | 0.86 | <3.4 | <0.4 | 10.9 |

NA = not available

4-t-octylphenol and 4-nonylphenol concentrations in primary effluents exhibited large variations representing differences of over an order of magnitude (Table 5-3). Elevated concentrations of 4-nonylphenol in primary effluent samples were usually associated with elevated concentrations of 4-t-octylphenol. 4-octylphenol was rarely seen in any samples (data not shown). Secondary effluent concentrations of 4-t-octylphenol for all facilities were below 180 ng/L. 4-nonylphenol in secondary treated effluents were significantly lower than the primary effluent concentrations and rarely exceeded 2,100 ng/L. Bisphenol A was found in concentrations between 300 and 700 ng/L in primary effluents and was below detection limits in all the secondary effluents.

Table 5-3. Occurrence of Phenolic Target Compounds in Primary and Secondary Treated Effluents in ng/L.

| | | 4-Nonyl phenol | 4-t-Octylphenol | Bisphenol A |
|---|---|---|---|---|
| Facility | Method | GC-NCI-MS | GC-NCI-MS | GC-NCI-MS |
| A | Prim. effluent | 12,420 | 1,165 | 664 |
| | Sec. effluent | 372 | <50 | <94 |
| C | Prim. effluent | 13,130 | 2,166 | 452 |
| | Sec. effluent | 2,224 | 74 | <94 |
| D | Prim. effluent | 1,799 | 432 | 274 |
| | Sec. effluent | 462 | 55 | <94 |
| E | Prim. effluent | 16,231 | 3,803 | 331 |
| | Sec. effluent | 1964 | 147 | <94 |
| F | Prim. effluent | 10,054 | 1,283 | 459 |
| | Sec. effluent | 1,317 | 179 | <94 |
| G | Prim. effluent | 27,567 | 2,063 | 311 |
| | Sec. effluent | 1,718 | 123 | <94 |
| H | Prim. effluent | 677 | 161 | 404 |
| | Sec. effluent 1 | 452 | 91 | <94 |
| | Sec. effluent 2 | 674 | 70 | <94 |

### 5.3.2 Potency of Estrogenic EDCs in Primary and Secondary Effluents

By using estrogenic potency factors (Table 3-2 for E-screen, Table 3-3 for ER-yeast), the presence of individual target compounds in effluent samples can be expressed in 17β-estradiol equivalent concentrations. The sum of target compounds expressed in EEq concentrations can be compared to the biological response in the E-screen as shown in Figure 5-1. Given the low estrogenic potency of testosterone, the HPLC-ELISA results only consider 17β-estradiol and estriol equivalent concentrations, while the GC-NCI-MS results consider the equivalent concentrations of all hormones (except testosterone) and phenolic compounds. For the E-screen, the total estrogenic activity in all but two primary effluents (Facilities A and D) was lower than the sum of individual target compounds based on GC-NCI-MS measurements, although results are in general agreement. With the exception of facility H, the ELISA results are much lower than predicted, based either on E-screen or GC-NCI-MS results. Taking the average proportion of activity from each compound across the seven influents, the total estrogenic activity was mainly caused by 17β-estradiol (21%) and estriol (62%), with rather small contributions from estrone (12%) and phenolic compounds (1.5%).

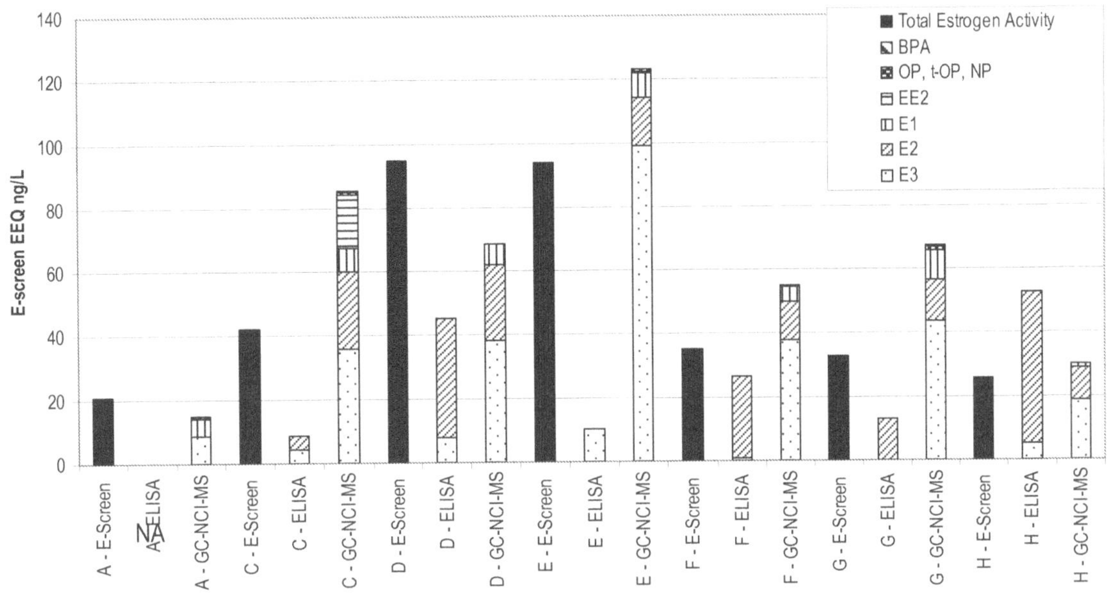


**Figure 5-1. E-screen Estrogenic Activity in Primary Effluents.**

The sum of target compounds expressed in EEq concentrations compared to the biological response in the ER-yeast assay is shown in Figure 5-2. For the ER-yeast, the total estrogenic activity in all but Facility D primary effluent was lower than the sum of individual target compounds based on GC-NCI-MS measurements, although again results are in general agreement. Taking the average proportion of activity from each compound across the seven influents, a different picture emerges. For the ER-yeast assay, the total estrogenic activity mainly consisted of 17β-estradiol (42%) and phenols (sum of NP and OP) (32%), with estrone (12%) and estriol (3.4%) making a smaller contribution.

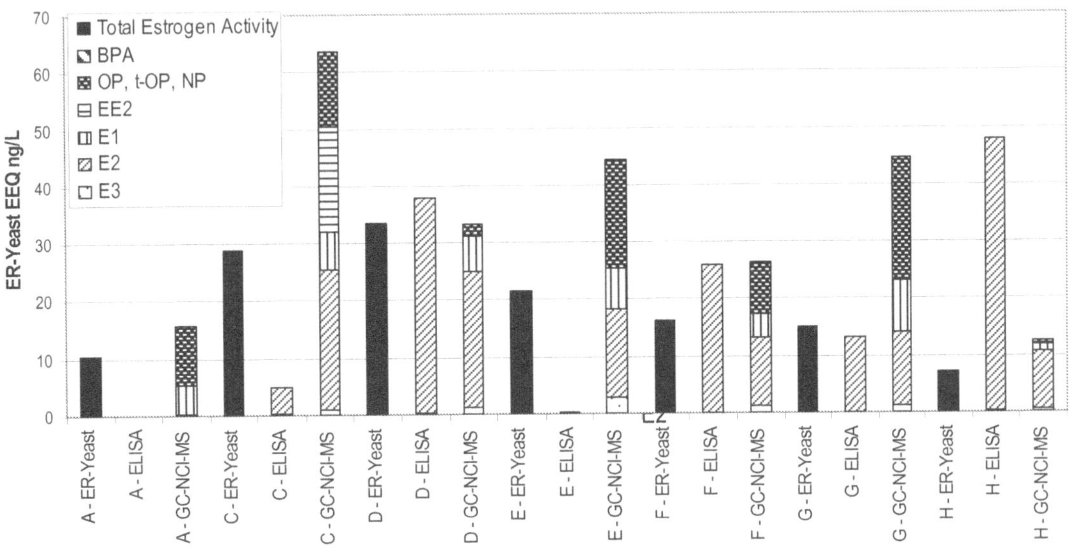


**Figure 5-2. ER-Yeast Estrogenic Activity in Primary Effluents.**

For secondary effluents (Figure 5-3), converting the E-screen potency of the GC-NCI-MS measurements (and averaging across the eight samples) indicates that the majority of the activity is from estrone (68%) and 17β-estradiol (22%). Although the estriol accounted for a large proportion of activity in the primary influents, in the secondary effluents estriol only accounted for 2.9%. In general, the E-screen results are in good agreement with the GC-NCI-MS results, although the activity is lower than predicted from the GC-NCI-MS results for five of the eight facilities. Although the ELISA was determined to have a lower detection limit than GC-NCI-MS, the results for the secondary treatment indicated two samples in which the 17β-estradiol was not detected in the ELISA but was with the GC-NCI-MS. Also, the ELISA result for one sample (Facility H) was markedly higher than either other analysis would indicate.

For secondary effluents (Figure 5-4), converting the ER-yeast potency of the GC-NCI-MS measurements (and averaging across the eight samples) indicates that the majority of the activity is from estrone (47%), nonyl- and octylphenols (27%), and 17β-estradiol (19%). With the yeast assay, estriol accounted for <0.1% of the activity in the secondary effluents. In general, the ER-yeast assay is in good agreement with the GC-NCI-MS results, although the yeast result for facility D seems unusually high.

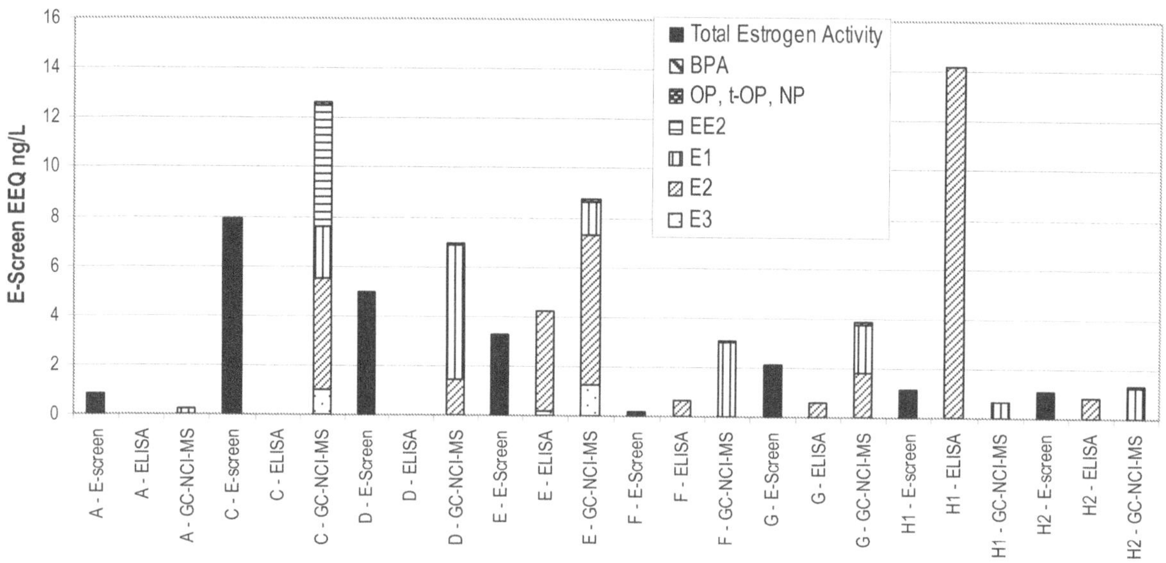


**Figure 5-3. E-screen Estrogenic Activty in Secondary Effluents.**

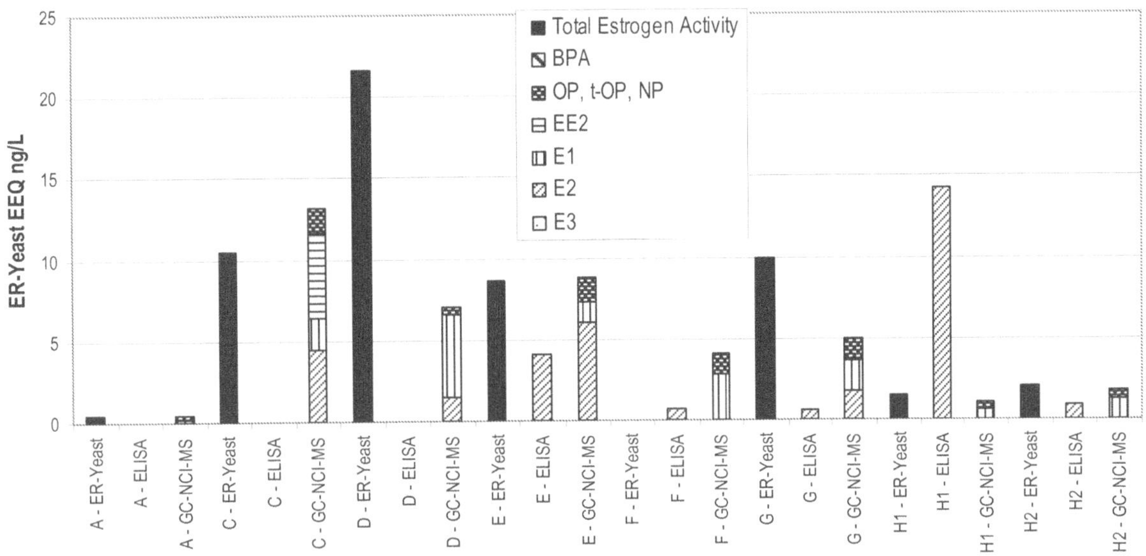


**Figure 5-4. ER-Yeast Estrogenic Activity in Secondary Effluents.**

### 5.3.3 Testosterone and Androgenic Activity in Primary and Secondary Effluents

Testosterone concentrations ranged from <3 to 57 ng/L based on GC-NCI-MS measurements in primary effluents (Table 5-4). Measurements of testosterone with ELISA consistently showed much higher levels, up to a 16-fold difference. The bioassay results were even higher than were the ELISA results. The TEq for the A-screen ranged from 190 to nearly 700 ng/L, and the AR-yeast results ranged from <90 to nearly 500 ng/L. These data strongly suggest that compounds other than testosterone are present in influents that contribute to the androgenic activity. Interestingly, the total androgen activity in primary effluents is quite a bit higher than the total estrogen activity.

With one exception (Facility H1), secondary treatment removed testosterone to concentrations below 5 ng/L based on HPLC-ELISA measurements, with two facilities detecting none in their treated effluents. Only two secondary effluents were above detection limit for the GC-NCI-MS. The bioassays also indicated good removal, with the A-screen results < 10 ng/L TEq for all samples, and one sample below detection limits. The AR-yeast assay has very poor sensitivity, and no secondary effluent samples were found active in that assay.

**Table 5-4. Occurrence of Androgen Activity and Testosterone in Primary and Secondary Treated Effluents in ng/L.**

| Facility | Method | Testosterone equivalents A-Screen | Testosterone equivalents AR-Yeast | Testosterone GC-NCI-MS | Testosterone HPLC-ELISA |
|---|---|---|---|---|---|
| A | Prim. effluent | 286 | <92 | <3.0 | NA |
| | Sec. Effluent | 1.84 | <92 | <3.0 | NA |
| C | Prim. effluent | 567 | 105 | 19.4 | 80.8 |
| | Sec. effluent | 2.37 | <90 | <3.0 | 1.9 |
| D | Prim. effluent | 676 | 382 | 21.7 | 79.1 |
| | Sec. effluent | <2.2 | <200 | <4.1 | <0.5 |
| E | Prim. effluent | 568 | 462 | 43.6 | 49.1 |
| | Sec. effluent | 1.76 | <90 | 1.3 | 4.1 |
| F | Prim. effluent | 597 | 488 | 24.5 | 45.2 |
| | Sec. effluent | 9.1 | <90 | <2.9 | <0.5 |
| G | Prim. effluent | 412 | <90 | 57 | 188 |
| | Sec. effluent | 1.64 | <90 | 4.9 | 2.4 |
| H | Prim. Effluent | 190 | 93 | 6.1 | 101 |
| | Sec. effluent 1 | 3.73 | <92 | <3.0 | 37.8 |
| | Sec. effluent 2 | 3.83 | <92 | <3.0 | 4.43 |

NA = not available.

### 5.3.4 BOD Loading and Estrogenic Activity in Primary Effluents

The organic loading to a secondary treatment process is usually expressed in BOD mass delivered per time. By considering all treatment facilities, an increased influent organic loading was related to an elevated influent estrogenic activity loading (Figure 5-5). The influent estrogenic activity loading, determined as the total estrogenic activity exhibited in the E-screen, showed a positive correlation with the BOD loading to wastewater treatment plants ($R^2 = 0.74$). 17β-estradiol and estriol (with ELISA measurements) exhibited a slightly stronger correlation ($R^2 = 0.77$) between the influent EEq loading and an increased BOD loading. This contribution confirms the dominant contribution of 17β-estradiol and estriol to the total activity in primary effluents.

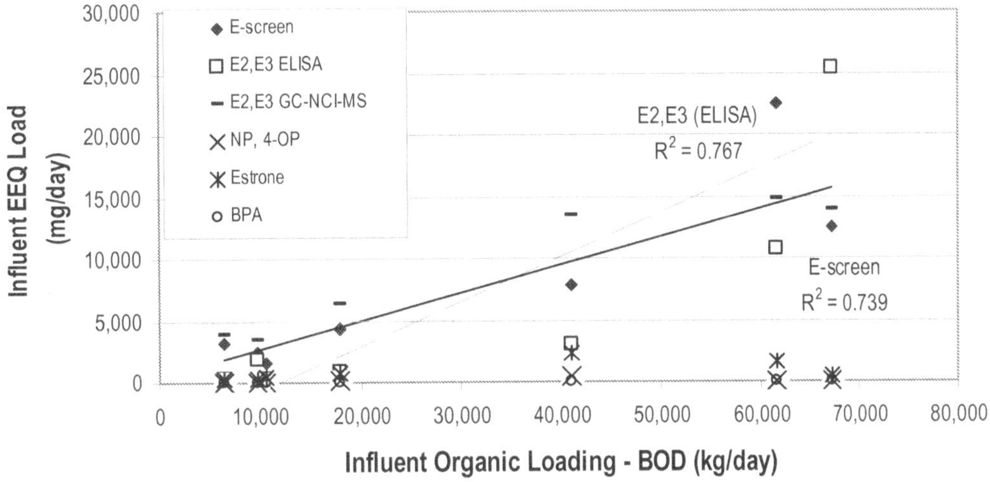


**Figure 5-5. Influent Organic Loading (as BOD5) vs. Influent E-screen EEq Loading for All Facilities.**

The relationship between the ER-yeast assay EEq and organic loading is not as clear (Figure 5-6), as the potency of estriol is much lower. Thus, although the relationship between the total estrogenic activity in the yeast assay was still somewhat strong, the relationship between 17β-estradiol and estriol and BOD was minimal ($R^2 = 0.14$). The low loading of nonyl- and octylphenol for the facilities with the highest BOD precluded any relationship between the phenols and BOD.

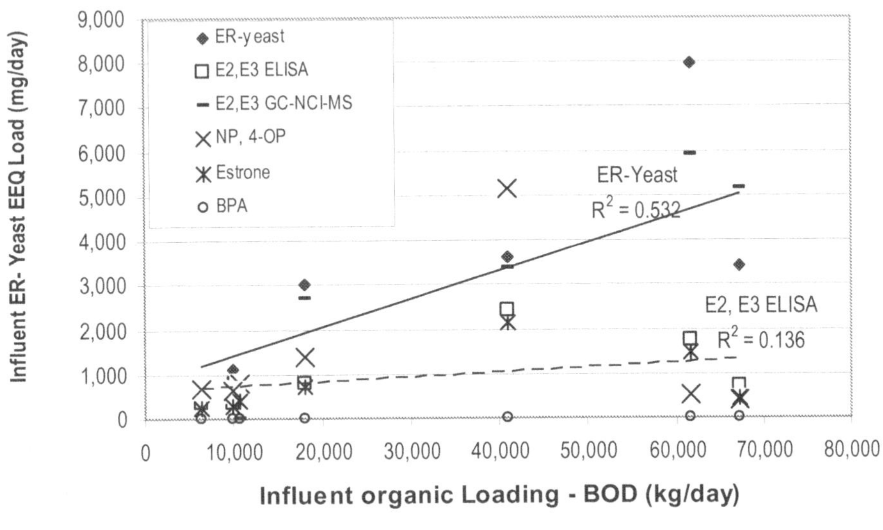


**Figure 5-6. Influent Organic Loading (as BOD5) vs. Influent ER-Yeast EEq Loading for All Facilities.**

### 5.3.5 Removal of EDCs During Secondary Treatment

During secondary treatment, the total estrogenic activity, as determined by the E-screen, was removed by 94% on average (Figure 5-7). Total estrogenic activity as determined by the ER-yeast was substantially lower, at only 63%, with high variability. Only one facility had measurable 17α-ethinylestradiol in the primary effluent, and 71% of this compound was removed by activated sludge treatment. 17β-estradiol was removed by 80% and 90%, as determined by GC-NCI-MS and ELISA, respectively. Estriol was removed below detection limits in the majority of the facilities and by >97% for the remaining facilities. Estrone was the most poorly removed of the compounds tested, with an average of only 48% removal. The poor removal is likely partially explained by the additional estrone contributions from 17β-estradiol oxidation. 4-t-octylphenol and 4-nonylphenol generally exhibited lower removal efficiencies as compared to the steroid hormones, and BPA results indicated removal of greater than 90%.

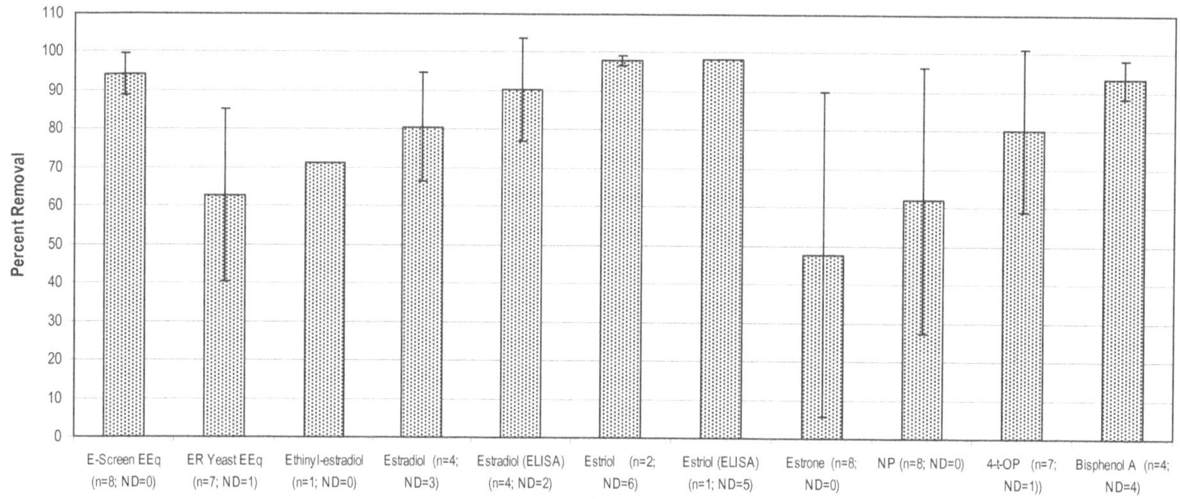


**Figure 5-7. Average Removal Efficiencies of Estrogenic Activity and Target Compounds During Treatment with Activated Sludge.** Values represent means ± s.d. with n = number of samples included in the average and ND = number of facility samples not included in the average because concentrations in secondary effluent were below detection limits. Unless otherwise indicated, results for target chemicals are from GC-NCI-MS. NP = nonylphenol, 4-t-OP = 4-t-octylphenol.

Removal of androgen activity, as indicated by the A-screen, was nearly 99%, with one sample removed to below the detection limit (Figure 5-8). The chemical analyses also indicated good removal, although because samples below the detection limit were not included in the average, the percent removal for GC-NCI-MS was greater than indicated in Figure 5-8. Another reason the bioassay results showed high percent removals was because the influents contained higher levels of activity than levels of testosterone. The removal for testosterone appears to be similar to the other estrogenic steroid hormones such as 17β-estradiol and estriol.

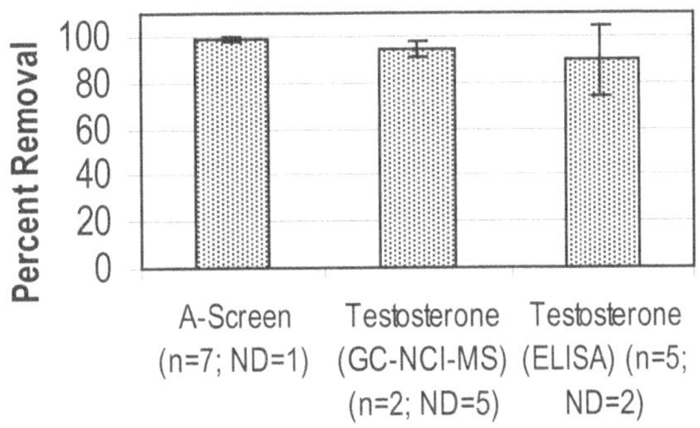


**Figure 5-8. Average Removal Efficiencies of Androgen Activity and Testosterone During Treatment with Activated Sludge.** Values represent means ± s.d. with n = number of samples included in the average and ND = number of facility samples not included in the average because concentrations in secondary effluent were below detection limits.

### 5.3.6 Relationship Between Operational Parameters and EDC Removal

Previous researchers (Kreuzinger et al., 2004; Clara et al., 2005) have reported that increasing the SRT results in an improved removal of EDCs. The present study covered a range of SRTs between 1.7 and 10 days. The data varied widely, especially for estrone, the ER-yeast assay, and phenolic compounds (Figures 5-9, 5-10). Because estrone is formed from the oxidation of 17β-estradiol, the removal is also confounded by the increased generation of the compound during treatment. Additionally, estriol is also formed from estrone. For that reason it can be useful to assess the relationship between the total estrogens and the retention times. The removal of the total estrogen load did appear to be improved by combining all the estrogen metabolites, although because estriol concentration levels were substantially higher in the primary effluents and removed to below detection limits in most cases, overall removal of total estrogens was quite high. Similarly, if the alkylphenol ethoxylate precursors to nonylphenol and octylphenol had been measured and combined with the alklphenols, their addition would likely indicate a greater degree of removal than looking at nonyl- and octylphenol alone.

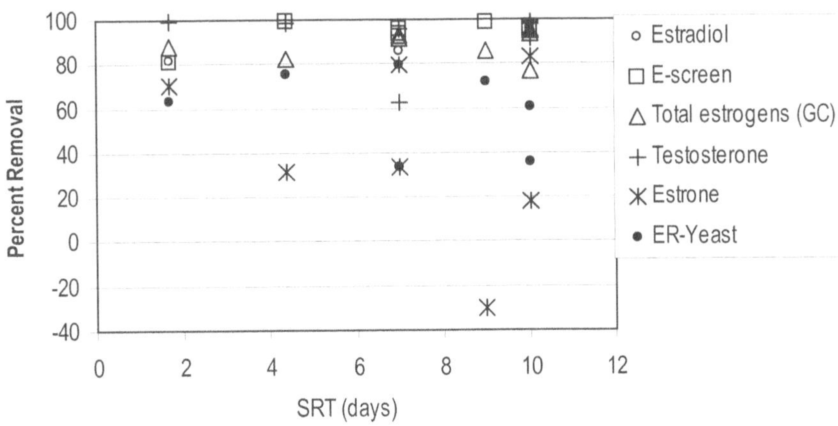


**Figure 5-9. Relationship Between SRT and Percent Removal for Hormones and Estrogenic Activity.**

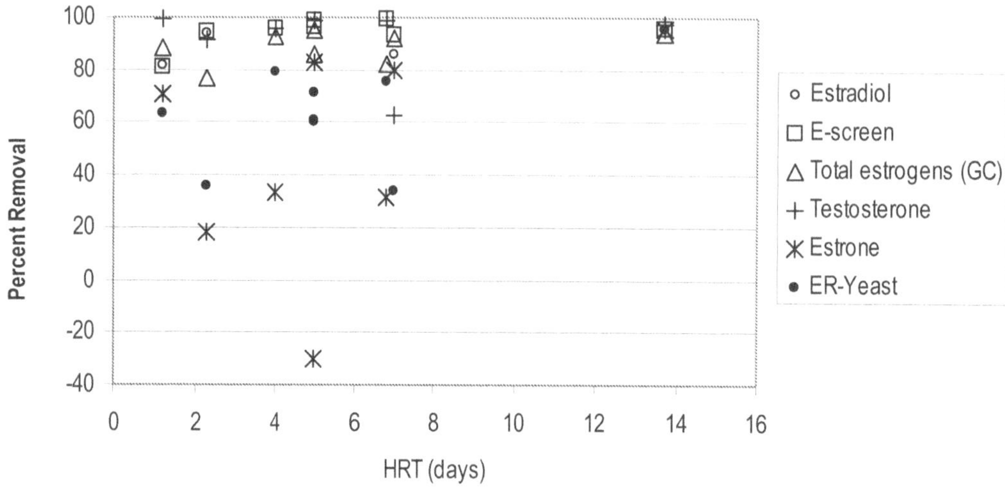


**Figure 5-10. Relationship Between SRT and Percent Removal for Phenols.**

HRTs were generally found to have a positive relationship with increased EDC removal. The HRT of Facility A was 13.7 hours, the longest in the study. This facility exhibited some of the greatest EDC removal (Figures 5-11, 5-12). The degree to which estrone and nonylphenol in particular were removed varied widely, especially at the lower HRTs, with some facilities exhibiting good removal at short HRTs.

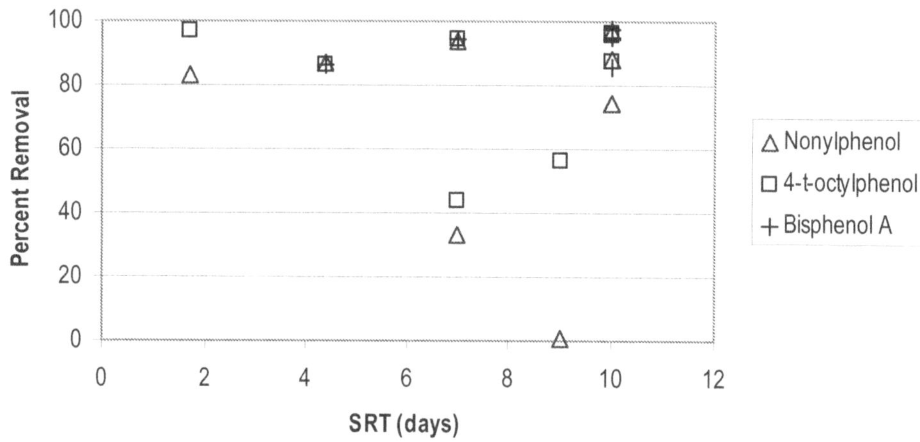


**Figure 5-11. Relationship Between HRT and Percent Removal for Hormones and Estrogenic Activity.**

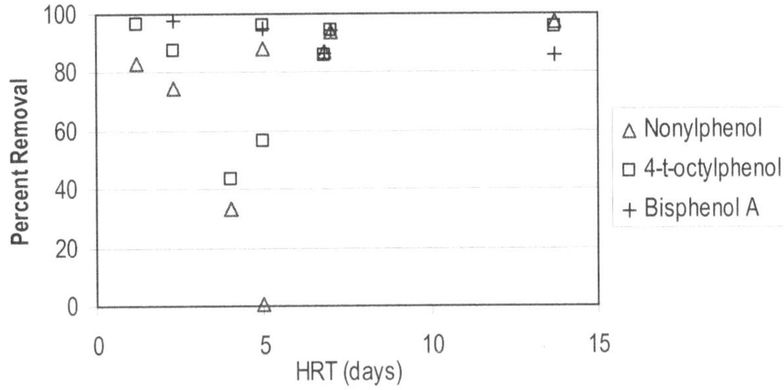


**Figure 5-12. Relationship Between HRT and Percent Removal for Phenols.**

The total estrogenic and androgenic activity removed during secondary treatment related well with BOD removal (Figure 5-13). These findings imply that within a range of BOD mass removal between 7,000 and 65,000 kg/day, secondary treatment processes are not limited in their ability to remove biological activity. It can also be concluded that the removal of bioassay activity follows mechanisms also responsible for BOD removal, namely biodegradation and adsorption to biosolids.

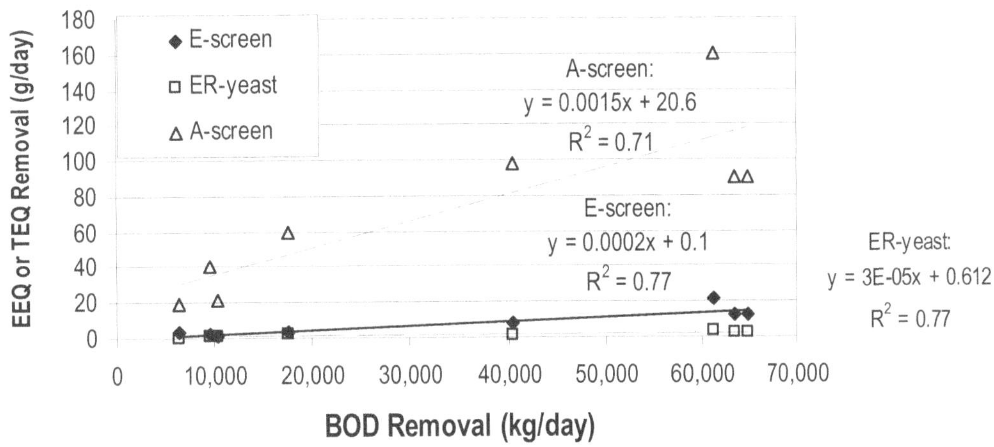


**Figure 5-13. Relationship Between BOD Removal and Bioassay Activity Removal.**

Kreuzinger et al. (2004) suggested that low food to microorganism ratios (F/M ratio) result in higher removal efficiencies of EDCs. To test this, the F/M ratio for each of the facilities was compared to the percent removal of the target compounds (Figure 5-14, 5-15). The facilities tested in this project had F/M ratios between 0.05 and 0.3 (kg BOD/kg MLSS x day). This range was fairly small, and all facilities within this range exhibited a similar degree of EDC removal.

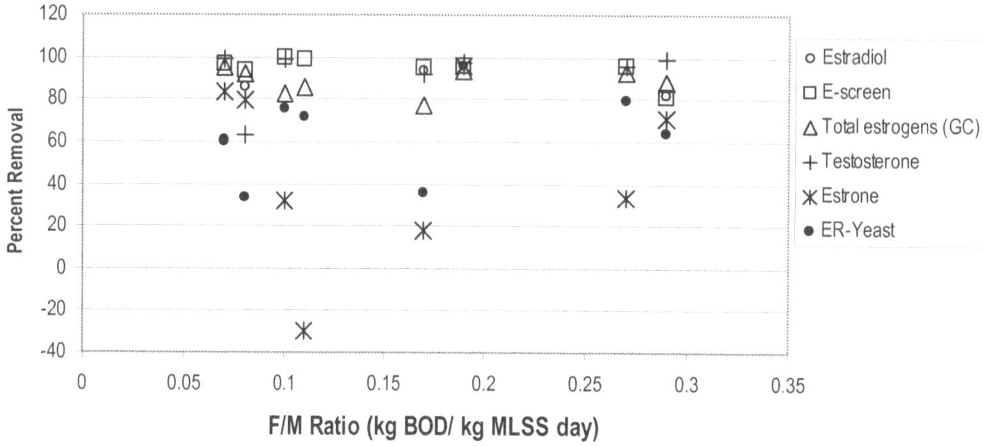


**Figure 5-14. Relationship Between F/M Ratio and Percent Removal of Hormones and Activity.**

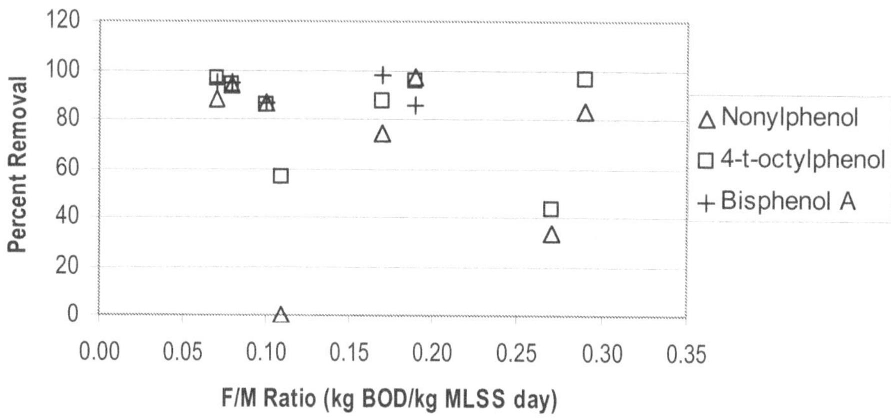


**Figure 5-15. Relationship Between F/M Ratio and Percent Removal of Phenols.**

## 5.3.7 UV and Chlorine Disinfection

### 5.3.7.1 Chlorine Disinfection

Composite samples were collected prior to and after chlorine disinfection at Facility H. This facility applied a chlorine dose of 3.5 mg/L with a contact time of approximately 45 minutes. Not all the target compounds were detected in the samples taken before disinfection. Of those detected, steroid hormones were removed by at least 64%, whereas phenolic compounds were reduced by 30%. In addition, estrogenic activity, as quantified by both the E-screen and ER-yeast assays, was reduced by the addition of chlorine. Westerhoff et al. (2005) conducted controlled chlorination studies with surface water, using dosages between 3.5 and 3.8 mg/L, and demonstrated a complete removal of steroid hormones, indicating that chlorination is efficient in reducing certain EDCs. Table 5-5 summarizes the concentrations detected before and after chlorine disinfection and the percent removal of the compounds. (Compounds and bioassays with no detections before disinfection have been omitted from the Table 5-5.)

Table 5-5. Removal of EDCs and Estrogenic Activity During Chlorination.

| Compound (Analysis) | Before Disinfection (ng/L) | After Disinfection (ng/L) | Percent Removal (%) |
|---|---|---|---|
| Nonylphenol (GC-NCI-MS) | 411 | 289 | 30 |
| 4-t-Octylphenol (GC-NCI-MS) | 26 | 18 | 30 |
| Estrone (GC-NCI-MS) | 2.7 | <0.5 | >82 |
| 17β-Estradiol (HPLC-ELISA) | 1.11 | <0.4 | >64 |
| E-screen (Bioassay) | 0.36 | 0.19 | 47 |
| ER-Yeast(Bioassay) | 1.11 | <1 | >10 |

### 5.3.7.2 UV Disinfection

UV disinfection is an emerging disinfection technology in water reclamation. To assess the effectiveness of UV disinfection in reducing the concentration of endocrine disrupting compounds and endocrine disrupting activity, samples were taken at two full-scale facilities prior to and after UV disinfection. The hydraulic retention time of the UV contact basin was taken into consideration when sampling so that the same slug of water was sampled. Facility E employed a Trojan 4000 medium pressure/high intensity system and Facility I employed an UltraTec low pressure/high intensity system. Figures 5-16 and 5-17 present concentrations of nonylphenol, 4-t-octylphenol, octylphenol, and bisphenol A for the samples taken prior to and after UV disinfection. For both full-scale UV systems, the results did not suggest any effect of UV disinfection on hormone or bioassay concentrations.

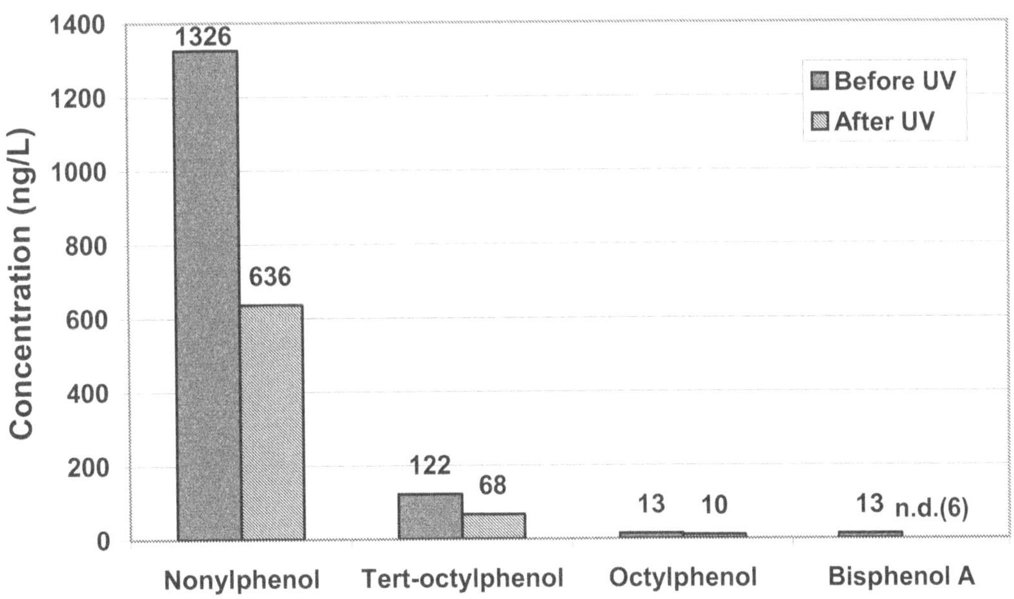


Figure 5-16. Concentration of Phenolic Compounds Before and After UV Disinfection at Facility E (Medium Pressure UV).

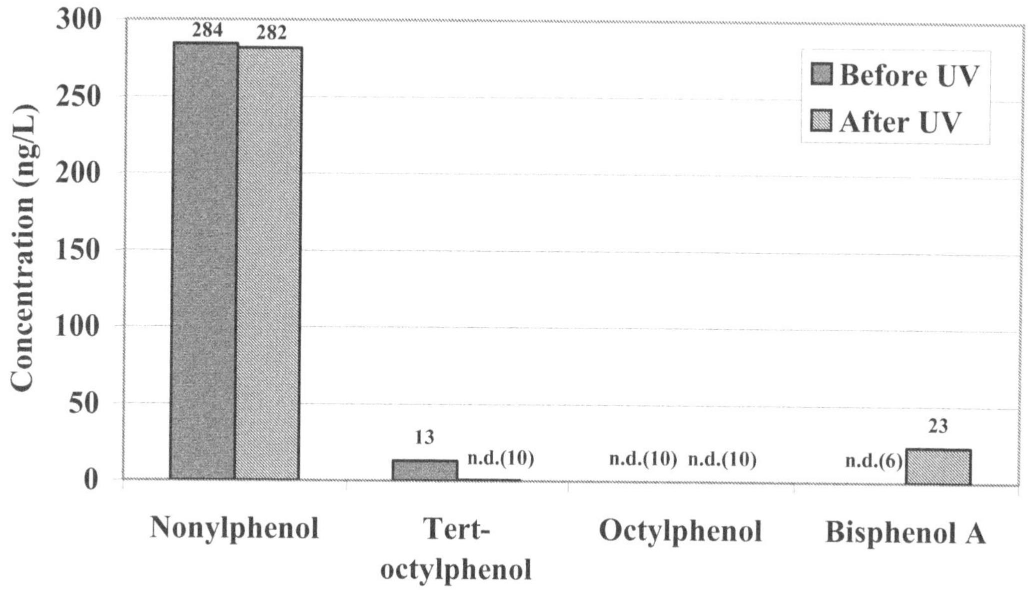


**Figure 5-17. Concentration of Phenolic Compounds Before and After UV Disinfection at Facility I (Low Pressure UV).**

Facility E demonstrated significant removal of the compounds nonylphenol and 4-t-octylphenol after exposure to medium range UV, reducing the concentrations by approximately half. Octylphenol and bisphenol A concentrations were also reduced, but concentrations for these compounds were very close to or below the detection limits of the GC-NCI-MS method and therefore the results are associated with some uncertainty. None of the compounds were removed during disinfection using low-pressure UV at Facility I.

## 5.4 Discussion

### 5.4.1 Occurrence of EDCs

Overall, concentrations of target chemicals found in this study are in line with values reported in the literature (summarized in Table 1-1). Only one facility had detectable 17α-ethinylestradiol in the influent sample that was slightly higher than reported in previous literature. That the remaining samples were all below 1 ng/L agrees with findings from other studies reporting similarly low concentrations for primary and secondary treated effluents (Ternes et al., 1999a; Belfroid et al., 1999; Snyder et al., 2001; Huang and Sedlak, 2001; Carballa et al., 2004; Vethaak et al., 2005). The remaining hormones were also in good agreement. As found in the present survey, estriol has also been reported in the highest concentrations in influents (Baronti et al., 2000; Solé et al., 2000), whereas estrone is generally found in the highest levels in the effluent. Concentrations of testosterone were in the same range as found by Lee et al. (2004). The phenolic compounds (4-nonylphenol, 4-t-octylphenol, bisphenol A) can vary markedly depending on the sources to the wastewater treatment plants (Lee et al., 2004). The researchers also found that concentrations varied widely among the facilities tested, with concentrations of nonylphenol > 4-t-octylphenol > bisphenol A.

The results of the bioassays also agree with results presented in the literature. The similarity of the ranges is fairly striking given the differences in assays used (ER-yeast, ER-CALUX) and global distribution. For example, in The Netherlands Vethaak et al. (2005), using the ER-CALUX bioassay, found influents ranging from 0.6-152 ng/L and secondary treated effluents <LOD to 2.6 ng/L EEq. In Sweden Svensen et al. (2003), using the ER-yeast assay, reported 3-29.8 ng/L EEq for influents and <0.1 – 12.1 ng/L after treatment with activated sludge. In the United Kingdom Kirk et al. (2002), using the ER-yeast assay, found 18-75 ng/L EEq for influents and up to about 10 ng/L in effluents. Murk et al. (2002) used the ER-CALUX and the ER-yeast assay and found that they resulted in very similar averages in influents (153 ng/L for ER-CALUX, 117 ng/L for the yeast) and 9.5 ng/L in the effluents with the ER-CALUX assay and 11.5 ng/L with the yeast. In the United States Holbrook et al. (2002) found up to 24 ng/L EEq in the influent and 10 ng/L after secondary treatment using the yeast assay. In Japan Tanaka et al. (2001) found influents averaged about 140 ng/L; after activated sludge effluents they averaged below 20 ng/L. For influents, results derived during this study ranged from 21 to 95 (E-screen) and 7 to 34 (ER-yeast) ng/L EEq; after secondary treatment the findings ranged from 0.18 to 7.9 (E-screen) and <0.5 to 21.6 ng/L EEq (ER-yeast).

With respect to androgens, the findings of higher androgenicity than estrogenicity in primary treated effluents is also corroborated by Kirk et al. (2002), Thomas et al. (2002), and Svensen and Allard (2004), who reported finding 4333, 635, and 211 ng/L dihydrotestosterone (DHT), respectively. The high detection limits for Thomas et al. of 23 ng/L and Kirk et al. of 113 ng/L precluded finding any activity after secondary treatment. Svensen and Allard (2004) achieved the low detection limit of 0.5 ng/L and found activity in secondary treatment in the low ng/L levels. For comparison, testosterone is about 0.28 times as potent as DHT in the A-screen.

### 5.4.2 Removal of EDCs During Secondary Treatment

Removal patterns are consistent with what other researchers have found (as reviewed in Johnson and Sumpter, 2001). Overall, the E-screen and A-screen indicate that activated sludge treatment does a very good job of removing biological activity. Because only one facility in the study had measurable levels of 17α-ethinylestradiol (other samples were below the detection limit of 0.7 ng/L), strong conclusions cannot be drawn regarding removal. Other researchers have pointed to 17α-ethinylestradiol as being fairly resistant to biodegradation (Ternes et al., 1999b; Weber et al., 2005) and likely to be removed via sorption to particles (e.g., Clara et al., 2004). 17β-estradiol and estriol are usually well removed, but in contrast to those two hormones, estrone removal varies markedly (Onda et al., 2003). As previously mentioned, estrone is formed by oxidation of 17β-estradiol (Ternes et al., 1999b), which may account for some of the poor removal. Removal of nonylphenol and 4-t-octylphenol was also quite variable. These are both breakdown products of alkylphenol ethoxylates, which can be formed during secondary treatment (Langford et al., 2005).

### 5.4.3 Potency and Relevance of EDCs

One interesting finding in this work was the different pattern of potency between the yeast and E-screen. Estriol has widely been disregarded as not important because of the low potency in the yeast assay. However, in the E-screen, estriol is only surpassed in potency by 17α-ethinylestradiol and 17β-estradiol. Because of the high concentrations of estriol in urine and high potency in the E-screen, estriol makes up a substantial proportion of the estrogenic activity found in primary treated effluents. Conversely, nonylphenol and octylphenol are less potent in the E-screen by about an order of magnitude than they are in the ER-yeast assay. Thus, according to the ER-yeast assay, the alkylphenols appear to be relevant contributors to estrogenicity in both the primary and secondary treated effluent.

Good removal notwithstanding, reports of intersex fish have appeared in Europe (Jobling et al., 1998; Vethaak et al., 2005) and recently in the Potomac River (*The Washington Post*, October 15, 2004), South Platte River (*The Denver Post*, October 3, 2004), the St. Lawrence River (Aravindakshan et al., 2004), and Lake Ontario (Kavanagh et al., 2004) in North America. Other biomarkers of estrogenicity such as elevated vitellogenin in male fish have also been widely reported. There is no consensus regarding what specifically is responsible for the feminization exhibited in fish. Although in vitro tests are mechanistically specific in their ability to identify estrogen receptor agonists, the complications involved in an aquatic organism (bioavailability, uptake, metabolism, excretion, interspecific variability) make the in vitro tests imperfect predictors of estrogenic potency in vivo (Hugget et al., 2003).

17α-ethinylestradiol is the most potent of the estrogens tested and is considered a likely candidate to be responsible for the estrogenicity in fish (Johnson and Sumpter, 2001; Vethaak et al., 2005). Metcalf et al. (2001) reported that medaka exposed to 17α-ethinylestradiol at concentrations as low as 0.1 ng/L developed intersex (testis containing ova), although this only occurred in one male of 33. At 100 ng/L, 90% of the fish were female and the remaining 10% of males all exhibited intersex. Full life-cycle exposure of zebrafish to 3 ng/L resulted in an all-ovary population of fish and in a complete inhibition of reproduction (Fenske et al., 2005). Van den Belt et al. (2003) reviewed the vitellogenin responses for many of the estrogenic compounds in a wide variety of fish over a wide variety of exposure conditions. 17α-ethinylestradiol caused vitellogenin induction at the lowest concentrations of any chemical tested, in all cases at levels $\leq$ 10 ng/L. Exposure of fathead minnows to 17α-ethinylestradiol at levels below 10 ng/L also altered the behavior in a male, reducing his ability to acquire a nest site (Majewski et al., 2002). Thus, there is adequate in vivo evidence that exposure to environmentally relevant levels of 17α-ethinylestradiol can cause the estrogenic effects found in wild populations.

The levels of the naturally occurring hormones required to exert estrogenic effects are fairly well characterized for 17β-estradiol in a wide variety of aquatic species, but data regarding the effects of estrone and estriol on aquatic organism are limited. A medaka full life-cycle test indicated that fecundity and fertility were significantly reduced at 28 ng/L 17β-estradiol, accompanied by an increase in intersex in males (Seki et al., 2005). According to the summary by Van den Belt et al. (2003), the lowest 17β-estradiol concentration to cause vitellogenin induction in males ranged from 20 to100 ng/L. With respect to estrone, Metcalfe et al. (2001) reported that intersex was found in two of 38 males at levels of 10 ng/L. At 1,000 ng/L, 92% of the fish were female, with the remaining males exhibiting the intersex condition. Thorpe et al. (2003) found that the lowest concentration of estrone needed to induce vitellogenin above controls was 3.3 ng/L, and Panter et al. (1998) found 32 ng/L estrone was required to significantly induce vitellogenin. In the only recent study of fish exposure to estriol, Metcalfe et al. (2001) found unusual results: a skewed sex ratio towards males at 10 ng/L and intersex in males at 1,000 ng/L. One study on slider turtles found estriol to be more potent than either 17β-estradiol or estrone in sex reversing turtles incubated at temperatures normally resulting in males (Bergeron et al., 1999). Clearly, the effects of estriol on aquatic organisms need to be studied more thoroughly before estriol is dismissed as being unimportant. Six of the eight secondary effluents analyzed in the study contained estrone at concentrations above 10 ng/L, with the highest measuring 46 ng/L and the highest concentration of 17β-estradiol being 6 ng/L. Thorpe et al. (2003) did find that estrogens (17β-estradiol and 17α-ethinylestradiol) acted in an additive way with regards to vitellogenin induction; therefore, it seems likely that levels found in secondary treated effluent could cause estrogenic responses in exposed fish.

Folmar et al. (2002) suggested that the phenolics are markedly less potent in the in vitro assays than they are in vivo and thus should not be dismissed as irrelevant. For nonylphenol, toxicity, not just estrogenicity, is important; Dussault et al. (2005) found 30,000 ng/L was lethal to more than half the fish that were exposed at that concentration. According to the summary by Van den Belt et al. (2003), the lowest nonylphenol concentration inducing vitellogenin ranged from 4,000 to 500,000 ng/L, well above the concentrations found in the secondary treated effluents (up to 2,224 ng/L) in this study. Similarly with 4-t-octylphenol, Van den Belt et al. (2003) report the lowest concentration to induce vitellogenin ranged between 4,800 and 100,000 ng/L, whereas the highest concentration in a secondary treated effluent in the study was below 200 ng/L. Finally, concentrations of bisphenol A were lower in the secondary treated effluents (all < 94 ng/L) than expected to cause induction of vitellogenin in fish (> 1,000,000 ng/L, according to Van den Belt et al., 2003 and references therein). These data make it seem unlikely that phenols in the effluents in this study alone are capable of causing measurable estrogenic responses in fish. However, Brian et al. (2005) also found that several of these estrogenic compounds (17β-estradiol, 17α-ethinylestradiol, nonylphenol, octylphenol, and bisphenol A) responded additively. Thus, it seems likely that levels found in secondary treated effluent contribute to the total estrogenic responses in exposed fish.

## 5.4.4 Relationship Between Chemical Analyses and Biological Activity

### 5.4.4.1 Estrogens

One of the goals of this project was to determine whether bioassay and chemical results agree or whether there were other biologically active compounds that could not be accounted for by the chemicals analyzed. Overall, the agreement between the estrogenic activity in the bioassays and the chemical analyses was good. Given the range of uncertainty based on the recoveries, duplicate analyses, variation in biological responses in determining the potency of chemicals, and interferences by organic matter in the yeast assay, it would be unwarranted to speculate that other estrogenic compounds are present or that receptor antagonists are limiting the biological response. Therefore, the researchers do not see evidence that there are unknown compounds that are contributors to estrogenic biological activity.

### 5.4.4.2 Androgens

In contrast to the reasonable agreement between the estrogenic bioassay and chemical results, the bioassay responses predict much higher levels of testosterone in the primary treated effluent than was determined with either chemical analysis technique. The obvious explanation, that other androgenic compounds are present, is likely. Thomas et al. (2002) did a toxicity identification analysis-based approach to evaluate which compounds were androgenic in wastewater extracts using an AR-yeast assay. Surprisingly, they found no testosterone but did find that many metabolites, including 4-androstenedione, 5α-androstanedione, 1-dehydrotestosterone and androsterone, contributed to the total activity. Based on this information the team obtained several standards (4-androstene-3,17-dione, 5α-androstane-3,17-dione, androsterone, 11β-hydroxyetiocholanolone) and derivatized them as with the other hormones to determine target ions and retention times. None of the compounds was seen using the hepta derivatives; therefore, the androgenic standards were analyzed using the pentaflourobenzoyl chloride derivative. Once compound profiles were set, the standard solutions were diluted into a series of mixes. A calibration curve was made, and all previously run samples were scanned for the additional hormones. The additional hormones would need to have been at concentrations above 200 ng/L to be detected with this approach. Only one of the compounds was seen, androsterone, in the runs from three facilities, with an estimate of the concentrations ranging from 1,900 to 6,300 ng/L. The researchers determined that the potency of androsterone relative to testosterone in the A-screen was 0.0025 and would only add 5 to 15 ng/L of activity. Although the concentrations of androsterone seem unusually high relative to other hormones, Kirk et al. (2002) referenced a 1970 study in which androsterone was found in raw sewage at levels between 5,700 and 14,040 ng/L.

As to whether androgens are having environmental impacts similar to the estrogenic effects seen in fish, no reports of domestic wastewater treatment plant effluents indicating evidence of excessive androgen exposure have been published, despite data indicating that final biological activities for both estrogenic and androgenic bioassays are similar in secondary treatment (generally < 10 ng/L EEq and TEq). It is unclear whether androgens are simply not causing any adverse effects or whether sensitive enough biomarkers of exposure have not yet been identified. One explanation for a lack of adverse androgenic effects in domestic wastewater is that unlike synthetic estrogens, which are widely used for hormone replacement therapy and as birth control, synthetic androgens are controlled substances and are not as widely used. Another explanation is that the androgen receptor does not appear as promiscuous as the estrogen receptor. One instance in which synthetic androgenic steroids are widely used is in cattle feed operations. Orlando et al. (2004) documented that male fathead minnows collected downstream of a large cattle feedlot exhibited signs of demasculinization: reduced interocular distance, decreased testis size, and reduced testosterone synthesis. They theorized that the demasculinization was caused via negative feedback by a synthetic androgen (trenbolone) at the hypothalamic-pituitary-gonadal axis. Evidence of masculinization of fish has also been reported in fish downstream of pulp and paper mills (Larsson et al., 2000 and references therein).

### 5.4.5 Relationship Between Operational Parameters and EDC Removal

That removal for most of the chemicals was above 80% for the phenols at 10 days is consistent with the findings of Strenn et al. (2003) and Lee et al. (2003). Clara et al. (2005) suggested a critical value for SRT (corrected to 10°C) of 10 days for removal of bisphenol A and natural estrogens. However, some facilities with SRTs as low as two days were capable of achieving removal of phenolic compounds above 80% and hormone removal above 95%. This would suggest that an SRT of less than 10 days is sufficient to remove the target compounds.

## 5.5 Conclusions

- Secondary treatment provides good removal of endocrine disrupting compounds, as noted based on chemical data as well as the biological activity of the effluent.
- Removal appears most variable for nonylphenol, 4-t-octylphenol, and estrone, which can actually be created during activated sludge treatment.
- Differences in potency of target chemicals lead to different estimates of what components are responsible for the activity. Because of variability in responses of aquatic organisms,

no one screening tool perfectly characterizes endocrine activity. However, the in vitro screening tools appear to provide reasonable estimates of estrogenic activity.

- Estrogenic activity was explained by the results of chemical analyses, but androgen activity was not. Testosterone was assumed to account for most of the androgenic activity, but results suggest androgen metabolites may be present and important.
- In terms of operational treatment parameters, BOD removal correlated well with removal of activity. Differences in SRT and F/M ratio were likely too small to make any meaningful comparisons. Not enough facilities had sufficiently long HRTs to adequately assess the relationship between HRT and removal, but the one facility with the longest HRT did have the best removal.

CHAPTER 6.0

# CHANGES IN EDCS AND ACTIVITY AFTER TREATMENT WITH MEMBRANE BIOREACTORS (MBRS)

## 6.1 Introduction

Membrane bioreactor (MBR) treatment has gained a lot of interest in the water reclamation community and has been proposed as a promising removal technology for EDCs and endocrine disrupting activity. Recently, studies have begun to examine the efficiency of MBR systems in removing EDCs from wastewater (Snyder et al., 2004). In this study, one full-scale MBR facility and three pilot-scale MBR facilities were investigated.

### 6.1.1 Pilot MBR Studies

The pilot-scale MBRs (provided by Kubota and Zenon Environmental) employed HRTs of 5 and 3 hours and SRTs of 10 and 21 days, respectively. The Zenon pilot-scale system designed for a capacity of 189 $m^3$/day was fed with primary effluent to a series of two anoxic tanks and one aerobic tank, followed by a membrane tank containing 10 Zenon hollow fiber membrane cassettes. The nominal pore size of the membrane was 0.04 μm. The system was operated at a flux of 20 $L/m^2 * h$ during sampling. A flow schematic of the system can be found in Figure 6-1. The same primary effluent was fed to the Kubota MBR pilot-scale system, consisting of one bioreactor equipped with two membrane units. Each membrane unit contained 50 flat plate cartridges composed of fine porous membranes. The average pore size of the Kubota membrane is 0.4 μm. During sampling, the system was operated at a flux of 51 $L/m^2 * h$. A flow schematic of the Kubota MBR is found in Figure 6-2. Table 6-1 summarizes the operating conditions of both the pilot- and full-scale systems.

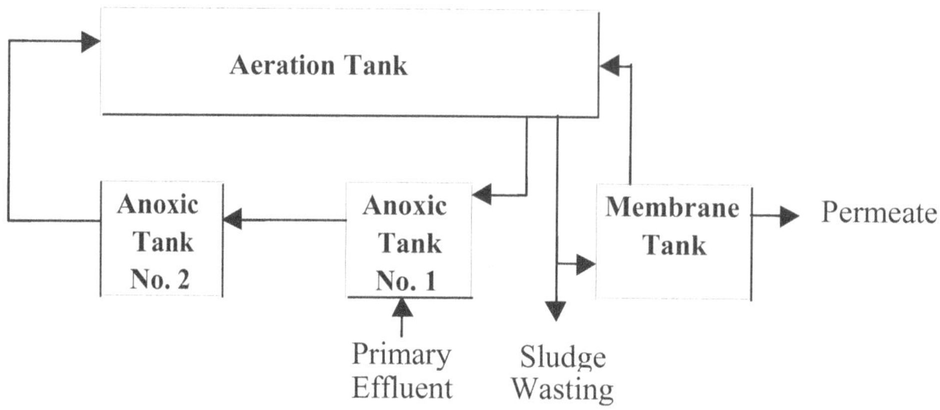

Figure 6-1. Flow Schematic of Pilot-Scale Zenon MBR Unit.

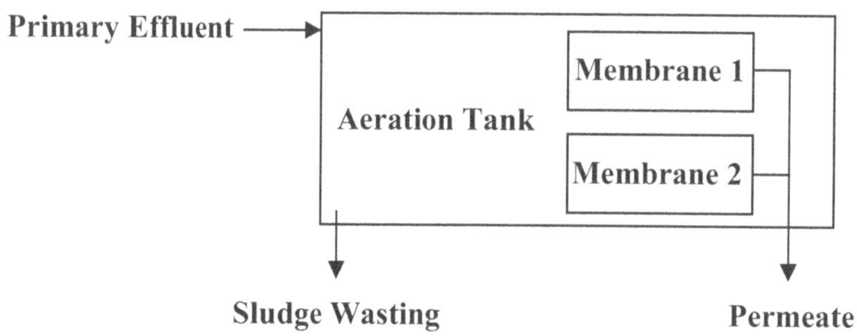

Figure 6-2. Flow Schematic of Pilot-Scale Kubota MBR Unit.

Table 6-1. MBR Operational Parameters.

| | Capacity | Hydraulic Retention Time (hours) | Sludge Retention Time (days) | Mixed Liquor Total Suspended Solids (mg/L) |
|---|---|---|---|---|
| Pilot Kubota | 0.014 MGD<br>53 $m^3/d$ | 5 | 10 | 10,000 |
| Pilot Zenon | 0.050 MGD<br>189 $m^3/d$ | 3 | 21 | 7,000 |
| Full-Scale Zenon | 2.4 MGD<br>9,072 $m^3/d$ | 5 | 15 | 14,000 |

### 6.1.2 Full-Scale MBR

The full-scale MBR in this study receives its wastewater from a highly industrial and commercial service area. This plant has a peak month design capacity of 9,072 $m^3/d$ (2.4 mgd) and utilizes Zenon Zeeweed© membrane bioreactors to provide secondary and tertiary treatment. Current monthly average influent flows are approximately 1.6 mgd. The nominal pore size of the membrane is 0.04 µm. Table 6-1 summarizes the operating conditions of the system. Tertiary effluent produced by the full-scale MBR is used for irrigation of a golf course and stream flow augmentation. A flow schematic of the full-scale MBR can be found in Figure 6-3.

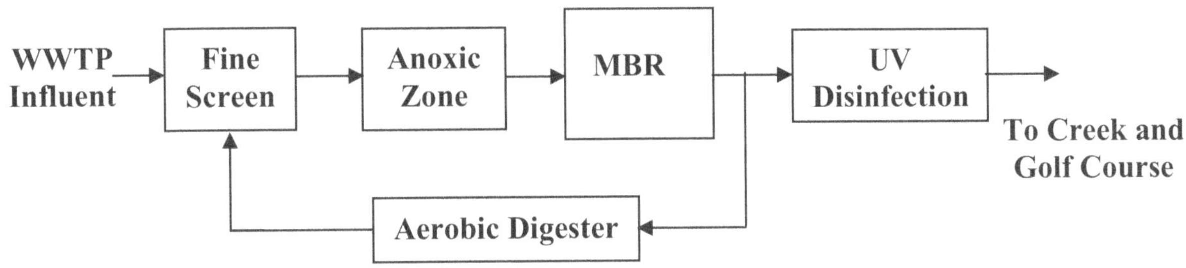


**Figure 6-3. Flow Schematic of Full-Scale Zenon MBR Facility.**

The wastewater characteristics gathered prior to sampling include five-day biochemical oxygen demand ($BOD_5$), total suspended solids (TSS), total Kjeldahl nitrogen (TKN), total phosphorus, alkalinity, and temperature. Discharge Monitoring Report data was collected for the full-scale facility from 2001 through 2004 to calculate influent flow, $BOD_5$, and TSS values. Laboratory data was also collected to calculate pH, temperature, and total phosphorus. Additional samples from June 2004 were also used to calculate TKN. Table 6-2 summarizes the collected influent wastewater characteristics and design values at the facility.

**Table 6-2. Full-Scale MBR Influent Wastewater Characteristics.**

| | | 2001 - 2004 | |
|---|---|---|---|
| Parameter | Unit | Average | Peak Month[1] |
| $BOD_5$ | mg/L | 231 | 261 |
| TSS | mg/L | 267 | 317 |
| TKN | mg/L | 41[2] | 43[2] |
| Alkalinity | mg/L | 230 - 380 | |
| Total Phosphorus | mg/L | 6.8 | 7.7 |
| | | Minimum Month | Maximum Month |
| Temperature | Deg C | 13.1 | 21.2 |

1. Peak Month concentrations were calculated by taking the maximum monthly mass loading divided by the peak month flow for the given year.
2. TKN data was only available for June 2004.

## 6.2 Fate of EDCs During MBR Treatment

Samples were taken at the influent and the effluent of the MBR systems. The fate of the target compounds was determined by both chemical measurements and bioassays. Table 6-3 summarizes the results of analyzing the samples with GC-NCI-MS, Table 6-4 summarizes the results of analyzing the samples with HPLC-ELISA, and Table 6-5 summarizes the results of analyzing the samples with the bioassays.

**Table 6-3. Influent and Effluent Concentrations (GC-NCI-MS).**

| | Full-Scale Zenon Influent | Full-Scale Zenon Effluent | Pilot-Scale MBR Influent | Pilot-Scale MBR Kubota Effluent | Pilot-Scale MBR Zenon Effluent |
|---|---|---|---|---|---|
| Nonylphenol | 5,715 | 284 | 1,544 | 615 | 400 |
| 4-t-Octylphenol | 417 | 12.9 | 465 | 128 | 30 |
| Bisphenol A | 597 | <47 | 350 | <15 | <15 |
| Testosterone | <0.5 | <0.5 | 143 | 1.6 | 1.6 |
| 17β-Estradiol | 25.3 | <0.5 | 21.7 | <0.5 | <0.5 |
| Estrone | 86.1 | 4.7 | 48.3 | 12.9 | 7.9 |
| Estriol | 169 | <0.5 | 249 | 3.9 | <0.5 |
| 17α-Ethinylestradiol | <0.5 | <0.5 | 10.3 | <0.5 | <0.5 |

Note: All concentrations in ng/L.

**Table 6-4. Influent and Effluent Concentrations (HPLC-ELISA).**

| | Full-Scale Zenon Influent | Full-Scale Zenon Effluent | Pilot-Scale MBR Influent | Pilot-Scale MBR Kubota Effluent | Pilot-Scale MBR Zenon Effluent |
|---|---|---|---|---|---|
| Estriol | 52.6 | <0.6 | 7.8 | <0.6 | <0.6 |
| 17β-Estradiol | 15.2 | <0.4 | 11.1 | <0.4 | <0.4 |
| Testosterone | 91.4 | 0.8 | 50.4 | <0.5 | <0.5 |

Note: All concentrations in ng/L.

**Table 6-5. Influent and Effluent Concentrations (Bioassay Results).**

| | Full-Scale Zenon Influent | Full-Scale Zenon Effluent | Pilot Scale MBR Influent | Pilot Scale MBR Kubota Effluent | Pilot Scale MBR Zenon Effluent |
|---|---|---|---|---|---|
| E-screen | 106 | 0.75 | 81.5 | 1.34 | 0.62 |
| ER-Yeast | 18.8 | 0.5 | 37.1 | 4.03 | 4.2 |
| A-screen | 670 | <5.4 | 543 | <3.3 | <2.2 |
| AR-Yeast | 110 | <92 | 246 | <200 | <200 |

Note: E-screen and ER-yeast concentration in ng/L EEq; A-screen and AR-yeast in ng/L TEq.

The phenolic compounds were greatly reduced for all samples collected from full- and pilot-scale MBRs. Nonylphenol had influent concentrations in the μg/L range and effluent concentrations from 284 to 615 ng/L. The relative removal varied between 60% and 90%. 4-t-octylphenol influent concentrations were reduced anywhere from 72% to 97% with the largest effluent concentration at 128 ng/L. Bisphenol A was reduced during treatment to below detection limit in all effluent samples.

17β-estradiol and 17α-ethinylestradiol, occurring in the influent at concentrations of 22 to 25 ng/L, and below detection limit and 10 ng/L, respectively, were both removed to levels below the detection limit for all systems. The removal of testosterone, estrone, and estriol was also high but not as complete. Estrone was still present in the permeate of MBR systems, which is consistent with findings reported by Snyder et al. (2004). That study examined two MBR systems that removed all other hormones tested to levels below detection, but one of the systems had a concentration of 8.43 ng/L estrone in the permeate of a pilot-scale MBR system.

Regarding estrogenicity of influent samples, the E-screen exhibited higher activities than observed in the ER-yeast assay. All MBR systems investigated, however, were capable of removing between 99% and 89% of estrogenic activity as measured by the two estrogen bioassays. An in vivo fish vitellogenin assay was conducted (see Appendix G). The tertiary treated effluent of the full-scale facility and MBR effluent from both systems were used, but none caused significant induction above controls. The A-screen concentrations were significantly higher than expected based on measurements of testosterone, suggesting that compounds other than testosterone were present in the influents and contributed to androgenic activity. For all MBR effluents, the androgenic activity measured with the two androgen bioassays was below the detection limit.

### 6.2.1 Comparison of Chemical Measurements to Bioassays

The chemical measurements (GC-NCI-MS and HPLC-ELISA) can be compared to the bioassays by taking the concentration of the compound and multiplying it by the potency factor for the compound specific to the bioassay. This determines the estrogen equivalent (EEq) for the compound. The sum of the EEqs as compared to the bioassays is presented in Table 6-6 and Figures 6-4 and 6-5. Potency factors are discussed in more detail in Chapter 3.0.

**Table 6-6. Bioassay Results and Relative Concentrations by Chemical Measures.**

| | Full-Scale Zenon Influent | Full-Scale Zenon Effluent | Pilot-Scale MBR Influent | Pilot-Scale MBR Kubota Effluent | Pilot-Scale MBR Zenon Effluent |
|---|---|---|---|---|---|
| E-screen | 106 | 0.75 | 81.5 | 1.34 | 0.62 |
| GC-NCI-MS EEq E-screen | 79.8 | 0.57 | 105 | 2.57 | 0.96 |
| ELISA EEq E-screen | 28.8 | 3.1 | 13.1 | 0.00 | 0.00 |
| ER-Yeast | 18.8 | 0.5 | 37.1 | 4.03 | 4.2 |
| GC-NCI-MS EEq ER-yeast | 40.6 | 0.72 | 44.0 | 2.14 | 1.19 |
| ELISA EEq ER-yeast | 15.5 | 3.1 | 11.1 | 0.00 | 0.00 |

Note: All concentrations in ng/L.

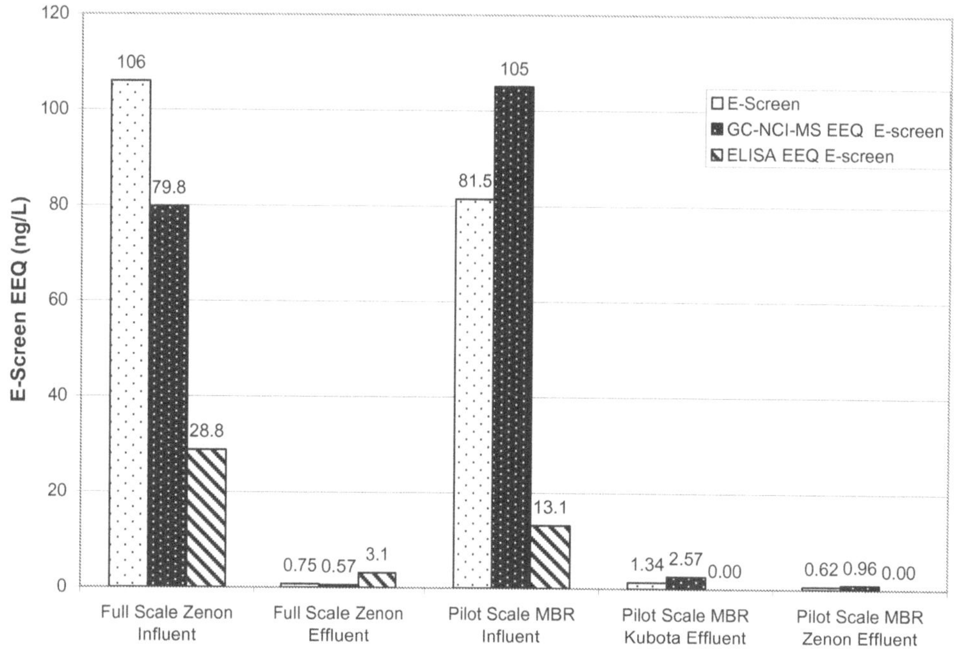


**Figure 6-4. Comparison of E-screen to Chemical Measurements using EEqs.**

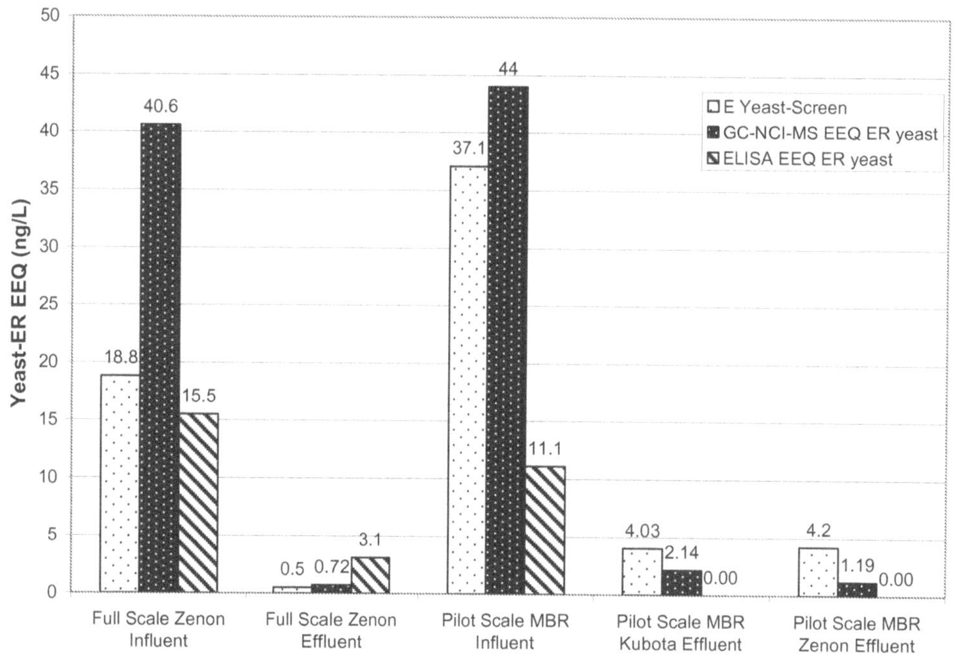


**Figure 6-5. Comparison of ER-Yeast to Chemical Measurements using EEqs.**

## 6.3 Comparison of Pilot-Scale MBRs to Full-Scale MBR

Removal of estrogen activity as determined by the E-screen was comparable for both the full-scale and pilot-scale systems. The influent and effluent concentrations for both types of systems are presented in Figure 6-6.

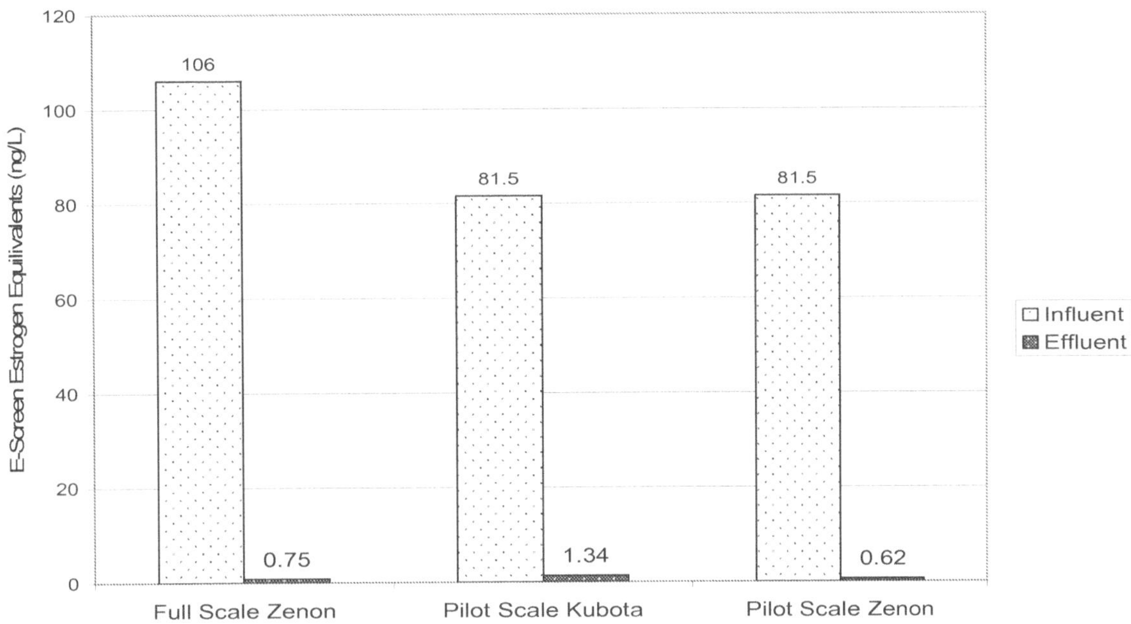


**Figure 6-6. Removal of E-screen EEq (ng/L) During MBR Treatment.**

Relative removal of nonylphenol and 4-t-octylphenol held constant between the full- and pilot-scale Zenon MBR, with nonylphenol removed less completely than 4-t-octylphenol in all cases. This may be caused by the effect of scaling down the system or less operating efficiency at the smaller scale. In addition, the Kubota pilot systems did not perform as well as the Zenon systems but achieved removal efficiency higher than 60% for the two phenolic compounds. Figure 6-7 provides a comparison of the removal efficiencies of the compounds for the phenolic compounds.

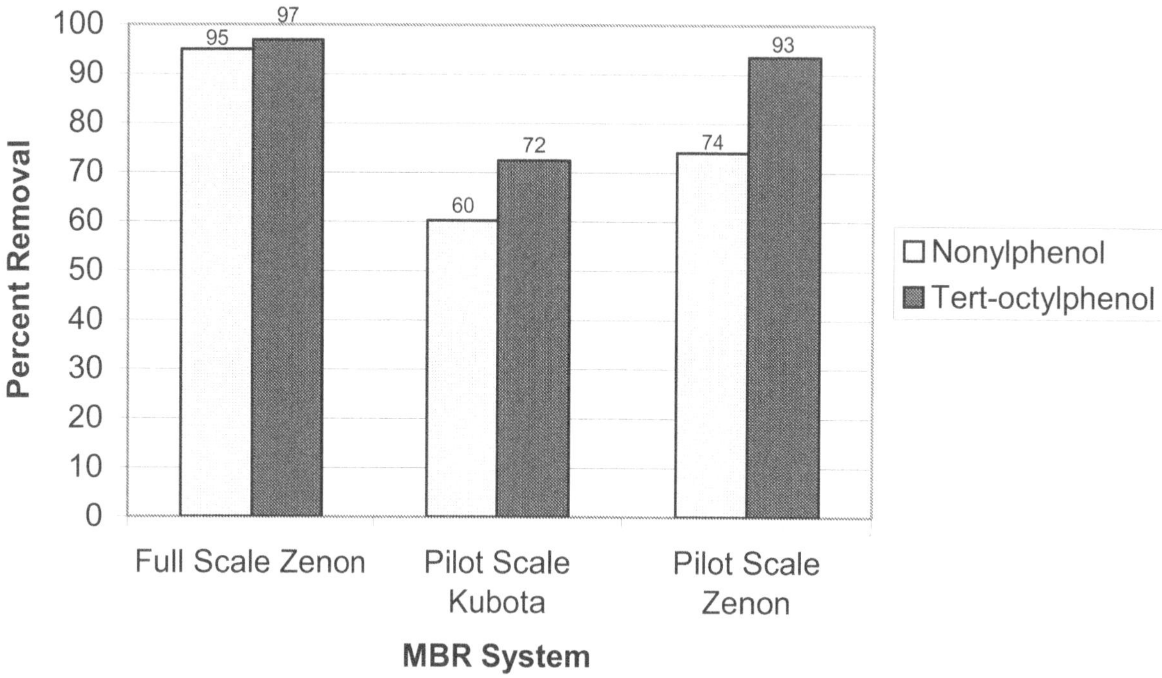


**Figure 6-7. Removal Efficiencies of Phenolic Compounds During MBR Treatment.**

## 6.4 Comparison of Full-Scale MBR to Activated Sludge Treatment

The results of the full-scale MBR facility were compared to the average performance of all of the facilities employing conventional activated sludge treatment (Chapter 5.0). Figure 6-8 represents the percent removal exhibited during full-scale MBR treatment and the average removal of facilities employing activated sludge.

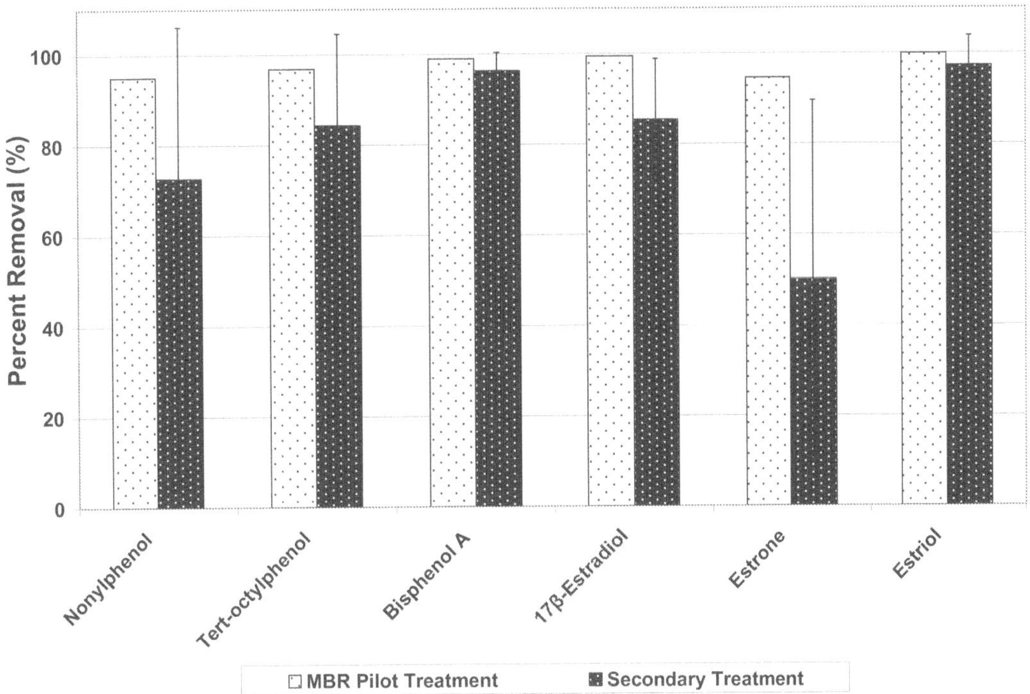


**Figure 6-8. Full-Scale MBR Compared to Average of Activated Sludge.**
Error bars represent the standard deviation of the secondary treatment samples.

The full-scale MBR system in this study appears to remove EDCs and endocrine disrupting activity to a level greater than secondary treatment employing conventional activated sludge. The largest difference in the removal of EDCs is for estrone. Estrone is a degradation product of 17β-estradiol and is generated during conventional secondary treatment. The better performance of MBR systems could be due to the fact that MLSS concentrations and SRTs for the MBR systems are typically greater than those of full-scale secondary treatment systems (see Tables 3-2 and 4-4). Both increased MLSS and SRT have been suggested to have a positive correlation with increased EDC removal (Lee et al., 2003; Clara et al., 2005). This study cannot draw definite conclusions about the relative removals of the two systems. Future research areas should include additional comparisons of MBR systems and conventional activated sludge systems regarding removal of estrogenic activity from wastewater.

# CHAPTER 7.0

# CHANGES OF EDCS AND ACTIVITY AFTER TREATMENT WITH HIGH-PRESSURE MEMBRANES

## 7.1 Introduction

Integrated membrane systems using microfiltration (MF) followed by reverse osmosis (RO) are the industry standard for indirect potable reuse applications via direct injection into an aquifer. The following key solute parameters have been identified by Bellona et al. (2004) in a comprehensive literature review to primarily affect solute rejection during nanofiltration and reverse osmosis: molecular weight (MW), molecular size (length and width), acid disassociation constant ($pK_a$), hydrophobicity/hydrophilicity (log $K_{ow}$), and diffusion coefficient ($D_p$). Key membrane properties affecting rejection include molecular weight cutoff (MWCO), pore size, surface charge (measured as zeta potential), hydrophobicity/hydrophilicity (measured as contact angle), and surface morphology (measured as roughness). Nghiem et al. (2004) conducted bench-scale experiments and reported that the adsorption and partitioning of steroid hormones led to a lower rejection of these compounds than what would be expected based on steric exclusion. Partitioning effects have been suggested by these researchers to result in detectable concentrations of hormones in permeates. In this study, one full-scale MF/RO facility and a laboratory-scale RO skid was examined to assess the removal of EDCs and biological activity during reverse osmosis treatment.

### 7.1.1 Pilot Studies

The rejection of EDCs was studied during an experiment using a one-stage membrane unit equipped with one 2540 TFC-HR (Koch Membrane Systems, San Diego, Calif.) spiral wound element (Figure 7-1). The TFC-HR is a reverse osmosis membrane commonly employed at full-scale installations practicing indirect potable reuse. Prior to the experiments, the membrane element was flushed with approximately 100 liters of Type II water (DI water).

Rejection studies were conducted in an effluent organic matter (EfOM) matrix representing both a common makeup of secondary treated effluents (after microfiltration) and a consistent background quality throughout all experiments. The EfOM was pre-isolated by collecting secondary effluent from a local wastewater treatment plant in Colorado. The secondary effluent was microfiltered and then concentrated to a 3:1 ratio using a XLE reverse osmosis membrane (DOW FilmTec). The concentrated effluent organic matter then underwent electrodialysis to reduce the conductivity to 1,500 microSiemens per centimeter, and the pH was adjusted to 2.5. To prepare the feedwater for the experiment, a certain amount of the EfOM concentrate was spiked into 190-L Type II water to obtain a nominal total organic carbon (TOC) concentration of 5 milligram per liter. The feedwater was adjusted to pH 6 using 3N NaOH solution, a conductivity of 934 µS/cm using sodium chloride, and a hardness of 120 mg/L as $CaCO_3$ using calcium sulfate. After a feed blank sample was taken as background, the target compounds were spiked from a stock solution to the feedwater in a stainless steel drum. Then the feedwater was mixed overnight by a pump bypass system. During RO operation, the recovery of the rejection test was established at 9.8%, representing typical hydrodynamic conditions of a single spiral wound element. After 20 L of feedwater passed the membrane vessel, initial samples were taken from the feed, concentrate, and permeate. After 180 L of feedwater passed the membrane vessel, final samples were taken from concentrate and permeate. After the experiment, the TFC-HR membrane element was flushed with a 0.01N NaOH solution for 10 minutes to remove potentially adsorbed organics and subsequently flushed with 100 L of Type II water. A permeate blank sample was taken after Type II cleaning to quantify a potential release of target compounds from the membrane surface.

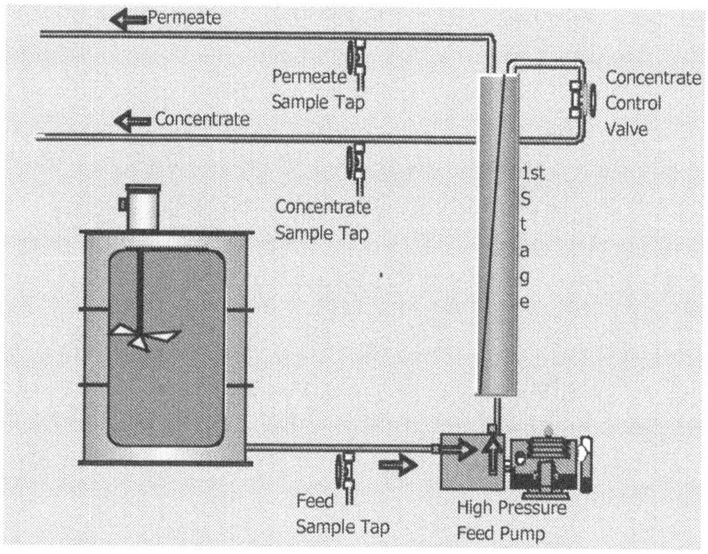


**Figure 7-1. Flow Schematic of Laboratory-Scale RO Skid.**

### 7.1.2 Full-Scale Integrated Membrane System (MF/RO)

Facility I is a full-scale MF/RO facility that currently operates at design capacity with a reverse osmosis treatment capacity of 45,000 $m^3$/d (12-MGD). Overall treatment at Facility I consists of grit removal, primary clarification, activated sludge treatment with nitrification/denitrification through a sequence of anoxic and aerobic zones, secondary clarification, tertiary dual-media filtration (anthracite, sand), disinfection using chloramines, microfiltration, and reverse osmosis. The MF/RO treated water is used for groundwater recharge via direct injection. Figure 7-2 represents the process schematic of the reverse osmosis treatment train at Facility I.

Full-scale testing occurred over a 48-hour time period using RO treatment Train #4 with a treatment capacity of 1.0 mgd (43.8 L/s). Each day, a set of grab samples was collected within 30 minutes from the feed (C1), concentrate of the second stage feeding the last stage (C2), and the permeate of each stage (P1, P2, and P3). The RO feedwater was nitrified/denitrified tertiary treated wastewater that was pH adjusted to 6.3 using sulfuric acid, microfiltered (US Filter, Warrendale, Pa.), and supplemented with anti-scalent (Hypersperse, GE Betz Inc.). The train operated at approximately 85% recovery, a specific flux of 0.07 gfd/psi (35.24 m/s kPa), and was configured as a three-stage array (24-10-5). The membrane employed in Train #4 was the RO membrane TFC-HR 8040 Magnum (Koch Membrane Systems, San Diego, Calif.). This is the same product also employed during the laboratory-scale experiments. Average bulk parameters representing the performance of Train #4 are summarized in Table 7-1.

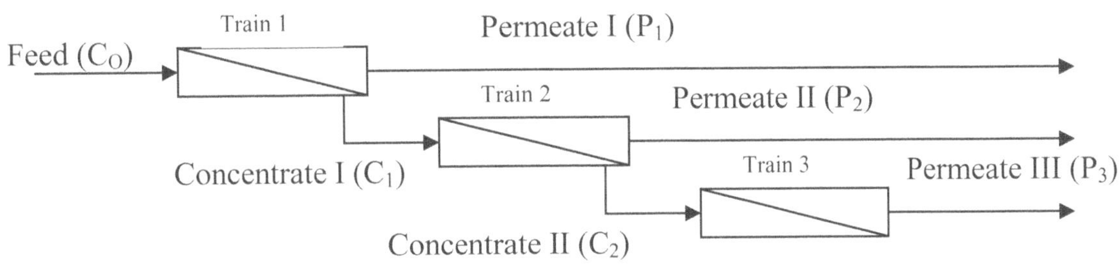


**Figure 7-2. Full-Scale Three-stage RO Train Configuration at Facility I.**

**Table 7-1. Average Bulk Parameters of Feed and Permeates for Full-Scale RO Train.**

| | Facility I | | |
|---|---|---|---|
| Sample | Conductivity (µS/cm) | UVA (1/m) | TOC (mg/L) |
| $C_0$ | 1,717 | 9.13 | 5.1 |
| $C_1$ | 3,748 | 21.3 | 13.8 |
| $C_2$ | 6,838 | 42.3 | 25 |
| $P_1$ | 38.1 | 0.17 | 0.15 |
| $P_2$ | 55 | 0.27 | 0.17 |
| $P_3$ | 96 | 0.23 | 0.24 |

## 7.2 Fate of EDCs During Laboratory-Scale MF/RO Treatment

The rejection of endocrine disrupting compounds from an EfOM water matrix during laboratory-scale testing resulted in high removal efficiencies for nonylphenol, 4-t-octylphenol and bisphenol A of 98.2%, 98.7%, and 99.4%, respectively (Table 7-2). It is noteworthy that the low concentrations observed in the permeate also occurred both in the initial permeate and the final blank after cleaning the system, which would suggest that either the Type II water carried some low concentrations of EDCs or they were released from the membrane itself. Bisphenol A is frequently used in the manufacturing process of nanofiltration and RO membranes. Steroid hormones 17β-estradiol, estrone, and 17α-ethinylestradiol were not detected in any of the RO permeate samples. Figure 7-3 represents the concentration changes of hormones observed during the experiment.

Table 7-2. EDC Concentrations During Laboratory-Scale RO Experiments.

| Compound | EfOM w/o Spike (ng/L) | Initial Feed (ng/L) | Initial Permeate (ng/L) | Final Concentrate (ng/L) | Final Permeate (ng/L) | Rejection (%) |
|---|---|---|---|---|---|---|
| Nonylphenol | 2,132 | 6,958 | 127 | 8,814 | 86.1 | 98.18 |
| 4-t-Octylphenol | 85.9 | 812 | 10.4 | 1,053 | 11.1 | 98.72 |
| Bisphenol A | 824.9 | 80,720 | 506 | 115,042 | 689.4 | 99.37 |
| 17β-Estradiol | n.d. | 21.9 | n.d. | 28.9 | n.d. | 100.00 |
| Estrone | 16.1 | 43.2 | n.d. | 57.5 | n.d. | 100.00 |
| 17α-Ethynylestradiol | n.d. | 25.0 | n.d. | 42.5 | n.d. | 100.00 |

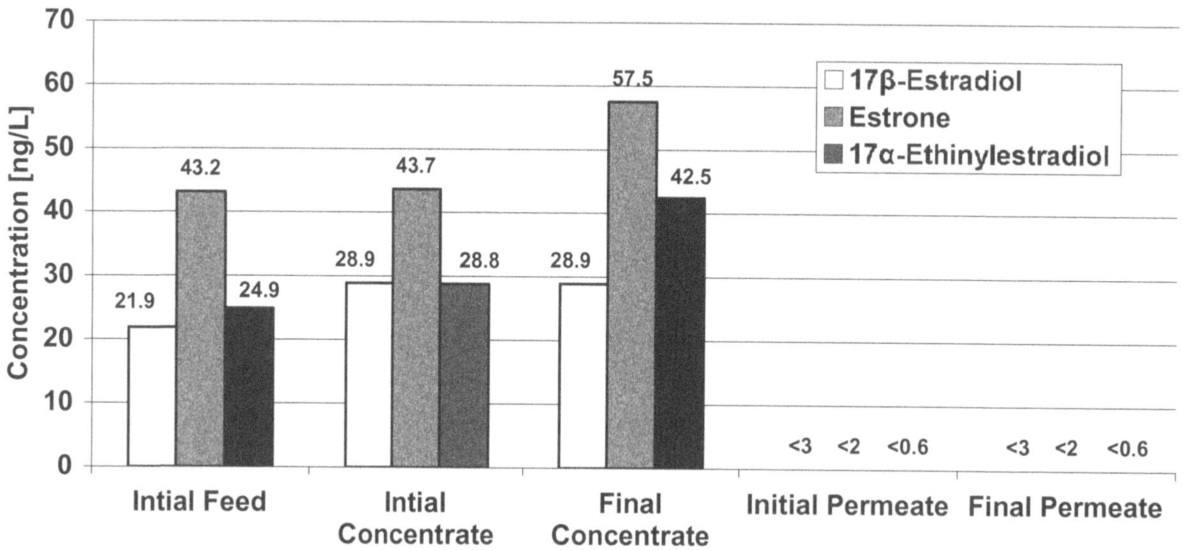


Figure 7-3. Concentration of Hormones During One-Stage TFC-HR Membrane Experiment.

## 7.3 Fate of EDCs During Full-Scale MF/RO Treatment

In the microfiltered tertiary treated effluent at Facility I, only the hormones estrone and testosterone were quantified. Sampling of the full-scale reverse osmosis treatment train demonstrated removal below the detection limit for the steroid hormones testosterone and estrone and nearly complete removal for nonylphenol and 4-t-octylphenol (Table 7-3). The concentrate concentrations increased through the treatment train, which is to be expected in a multi-stage reverse osmosis train (Figures 7-3, 7-4). No biological activity was quantified in any of the permeate samples.

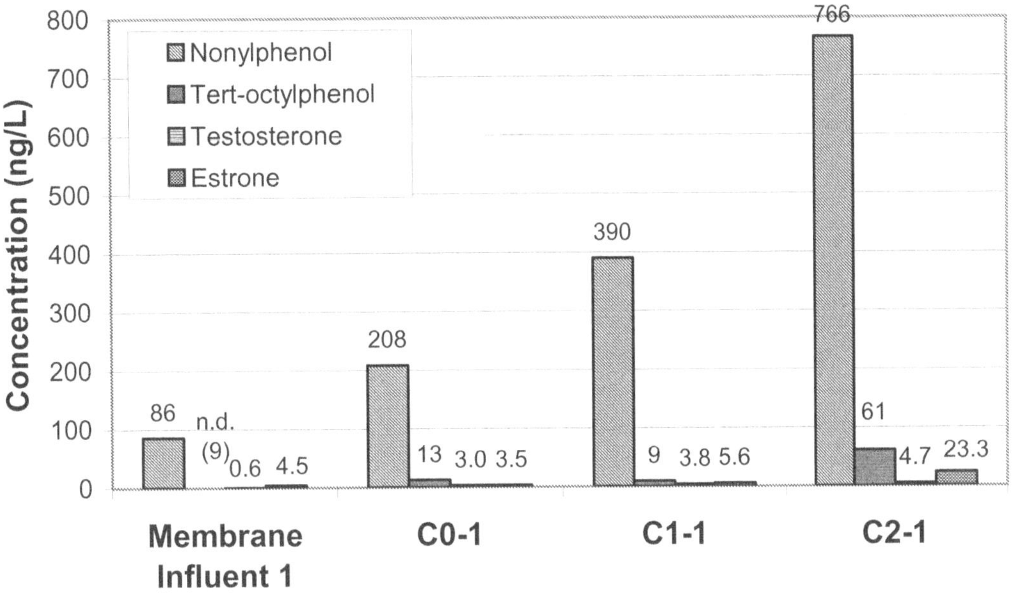


Figure 7-4. Day 1. Concentrate Concentrations of Nonylphenol, 4-t-Octylphenol, Testosterone, and Estrone by Full-Scale Reverse Osmosis.

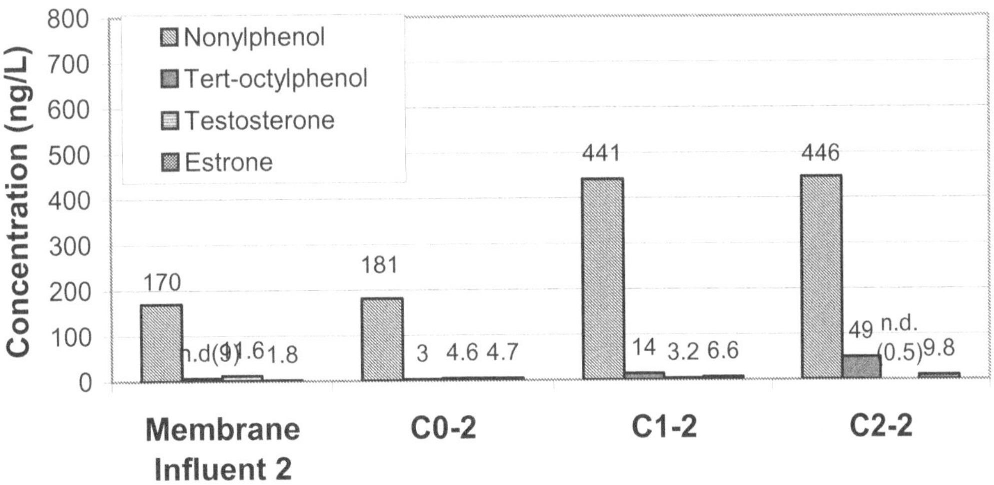


Figure 7-5. Day 2. Concentrate Concentrations of Nonyyphenol, 4-t-Octylphenol, Testosterone, and Estrone by Full-Scale Reverse Osmosis.

Table 7-3. Facility I Reverse Osmosis Concentrate Concentrations (in ng/L).

| Sample | P1-1 | P2-1 | P3-1 | P1-2 | P2-2 | P3-2 |
|---|---|---|---|---|---|---|
| Nonylphenol | <15 | 33 | <15 | 48 | 40 | 28 |
| 4-t-Octylphenol | <9 | <9 | <9 | 13 | <9 | <9 |
| Testosterone | <0.5 | <0.5 | <0.5 | <0.5 | <0.5 | <0.5 |
| Estrone | <0.5 | <0.5 | <0.5 | <0.5 | <0.5 | <0.5 |

The tested compounds in these experiments are all considered hydrophobic, with log $K_{ow}$ values in the range of 3.13-5.28. Their molecular weights vary from 266-340 Daltons and therefore represent a size well above the molecular weight cutoff of the TFC-HR membrane. Therefore, hydrophobic/hydrophobic interactions between solutes and membrane as well as steric exclusion are likely responsible for high removal efficiencies of EDCs. After chemical and hydraulic cleaning, nonylphenol, 4-t-octylphenol, and bisphenol A were detected in the cleaning water (DI) with concentrations close to or slightly higher than in the permeate. However, octylphenol and the hormones tested for were not detected in cleaning water. No steroid hormones and biological activity were, however, detected in laboratory- or full-scale permeate samples, indicating that reverse osmosis represents a very efficient barrier for EDCs. These findings are consistent with previous studies (Drewes et al., 2005).

# CHAPTER 8.0

# ASSAYS FOR DETERMINING ENDOCRINE DISRUPTION OF THE THYROID SYSTEM

## 8.1 Introduction

### 8.1.1 Background Information

Assessing evidence of endocrine disruption in the thyroid system is much more complicated than the relatively straightforward assessment of estrogen or androgen receptor agonists. The thyroid system is involved in growth and development and plays a key role in the regulation of metabolism. In brief, the thyroid hormone (thyroxin or T4) is secreted by the thyroid in response to signals from the pituitary. Thyroxin is the amino acid tyrosine bound to four iodine atoms, which is activated to its more potent form (T3). Virtually all cells in an organism are targets for thyroid hormone. The thyroid receptor works in a similar manner to the androgen and estrogen receptors; once T3 binds to a thyroid receptor, the complex binds to a thyroid response element on the DNA and elicits gene expression. Until recently, there were no reports of environmental chemicals that bound and activated the thyroid hormone receptor (DeVito et al., 1999). This has been updated with a report that some halogenated bisphenol A compounds (Kitamura et al., 2002) bound to the thyroid receptor and elicited a response in vitro.

### 8.1.2 Chemicals that Interfere with the Thyroid System

Howdeshell (2002) reviewed chemicals that interfere with the thyroid system and concluded that nearly every step in the synthesis, transport, metabolism, and elimination of thyroid hormone has been documented to be disrupted by synthetic chemicals. Examples include a large list of chemicals that alter the uptake of iodide by the thyroid gland (including 2,4-D, aldrin, PCBs), perchlorate (which competes with iodide at the sodium/iodide symporter), and PBBs, which alter the enzyme that iodinates tyrosine.

### 8.1.3 Summary of OECD Draft Detailed Review Paper on Thyroid Hormone Disruption Assays

Recently, the Organisation for Economic Co-operation and Development (OECD) released a draft detailed review paper on thyroid hormone disruption assays (OECD, 2005). Much of the paper describes in vivo mammalian assays which would not be especially relevant to wastewater samples, but it also reviews a variety of in vitro tests that might be useful in screening chemicals (or perhaps wastewater extracts), as well as endpoints for fish and frog in vivo assays.

#### 8.1.3.1 In Vitro Thyroid Assays

Receptor binding studies can be used to determine whether chemicals can displace T3 from the thyroid receptor (four isoforms of the thyroid receptor are known, and they appear to function as heterodimers). Analogous to the ER- and AR-yeast assays, yeast cells transformed with thyroid receptors have also been produced and can determine whether chemicals can activate thyroid receptor dependent transcription. An in vitro bioassay, $GH_3$ cell assay for thyroid hormone action, has been developed to detect chemicals that obstruct the ability of the thyroid receptor to respond appropriately. The assay uses a rat pituitary somatotroph cell line that responds to stimulation of thyroid hormone by proliferation. This assay is relatively new (1996) and has not been used with environmental samples. Another cell line, FRTL-5 (derived from rat thyroid tissue), was described as being able to test the effects of several types of thyroid toxicants, including iodide uptake inhibitors or other thyroid functions, although no other details are given.

#### 8.1.3.2 In Vivo Fish Assays

With respect to thyroid effects in fish, PCBs have been shown to alter T3 and T4 levels in several different fish species. Although several parameters of thyroid status can be measured in fish, such as thyroid hormone levels, receptor binding using fish thyroid receptors, or the deiodinase enzyme activity, there is not enough information to be able to put these results into context regarding the biological relevance of changes in the status of the fish thyroid system. Authors of the OECD report concluded that there were no current in vivo or in vitro assays that were developed thoroughly enough to be recommended as tests to screen for thyroid active compounds. One potentially useful thyroid endpoint in a fish is that thyroid-driven metamorphic changes in fish do occur during changes between larval and juvenile stages. In "asymmetrical" flatfishes, some external morphological changes occur that can be easily followed (eye migration). The authors of the report indicate this type of flatfish metamorphosis assay may be suitable, but because of the lack of availability of flatfish and work required to develop the assay more thoroughly, it can not be recommended at this time.

#### 8.1.3.3 In Vivo Frog Metamorphosis Assays

Amphibian metamorphosis assays are being developed for detection of thyroid active substances (OECD, 2004; Degitz et al., 2005). It is well known that the thyroid hormone drives the metamorphosis of tadpoles into a frog. The original EDSTAC recommended test, the tail resorption test, was a 14-day assay beginning exposures of *X. laevis* at Stage 60 (EDSTAC, 1998). It was determined this stage occurs concomitantly with the endogenous T4 peak, which was too late to sensitively alter responses with an exogenous chemical. In effect, the metamorphic progression had already been set in motion and could not easily be altered (Degitz et al., 2005). A more sensitive screening test is being developed (OECD, 2004), although most parameters have not been determined, including species, exposure window, and endpoints. Two of the most promising species are *X. laevis* and *X. tropicalis*. *X. laevis* would be the obvious candidate because it has been so commonly used in toxicity studies. The metamorphic progression has been extremely well-characterized by Neuwkoop and Faber (1994). However, it is genetically very complicated (it has an oliotetraploid genome), and it has a relatively long life cycle. Therefore, recently *X. tropicalis*, a closely related species with similar toxicity profiles, has been suggested for use. *X. tropicalis* is smaller, develops more rapidly (but corresponds well to the Neuwkoop and Faber progression), and has a diploid genome that is mapped. The drawbacks are that the husbandry has not been worked out as well and availability is not yet as good.

Regarding the most appropriate window of exposure, thyroid hormone dependent morphogenesis begins at Stage 54 and by Stage 60, many tissues have programming defined. The metamorphic climax (stage 60 to 66) is where the dramatic changes from tadpole to frog occur, including tail resorption. The EPA lab in Duluth, Minn. is experimenting with beginning at either Stage 50 or 54 and trying assay duration of 14-21 days (Degitz et al., 2005).

The OECD (2004) reviews several likely endpoints for an amphibian metamorphosis assay. The morphological endpoints include developmental stage and hind limb development. Histological endpoints include changes in the thyroid gland or the pituitary gland. Some important biochemical parameters are the measurement of the thyroid hormones in plasma or whole body homogenates and enzyme activity of iodothyronine deiodinase (converts T4 to T3 to activate). Many molecular biomarkers may be very promising in terms of sensitivity and short exposures. In many cases, mRNA is induced within hours of exposure to a chemical. Changes in mRNA expression can be quantified with single gene expression assays using real-time polymerase chain reaction (RT-PCR) which is robust, sensitive, and fairly cost effective. Some of the likely candidates would be mRNA that codes for thyroid receptor and stromelysin -3 (involved in cell migration and apoptosis). Using multiple gene expression assays, such as gene arrays, may help identify other biomarkers of thyroid system disturbance. Overall, it is clear that much work remains to be done before a standardized, validated method is recommended.

The goal for this project was to try using an assay similar to the one being developed by the U.S. EPA lab in Duluth. The researchers planned on using *X. tropicalis* because they had easy access to two different sources and decided using a smaller, more quickly developing organism would be beneficial for tests involving the collection of effluent. They used sodium perchlorate, a known inhibitor of the thyroid system that inhibits the uptake of iodide at the sodium-iodide symporter, as a positive control and also exposed the tadpoles to secondary treated effluent. The researchers started tadpoles at Stage 50-51, when the hind limbs are just starting to be formed. The endpoints they planned on were developmental stage and thyroid histology.

## 8.2 Methods

Two separate experiments were conducted. In the first experiment, the researchers obtained *X. tropicalis* at Stage 48 from Nasco (Ft. Atkinson, Wis.). Tadpoles were cultured at WSLH in dechlorinated Madison tap water at 26°C for seven days until they reached Stage 50-51. To initiate the experiment, tadpoles were classified using Nieuwkoop and Faber (NF) developmental stages. Ten tadpoles at NF Stage 50-51 were placed in 20 L aquaria with three aquaria per treatment and 8 L of the testing solution. Sodium perchlorate was used as the positive control at four concentrations: 30, 62, 250, and 1,000 ppb. Dechlorinated tap water was used as the control, and Facility A secondary effluent was also used to rear the tadpoles a testing solution. The aquaria were kept in a walk-in incubator at 26°C. The dissolved oxygen, temperature, and pH were checked daily. Seventy-five percent of the water was exchanged daily, and the tadpoles were fed Sera Micron three times per day. After 20 days, the tadpoles were euthanized, weighed, and preserved in 10% buffered formalin. The tadpoles were staged within three classes: below Stage 54, between 55-57, and 58 or above. The characteristics to distinguish the stage are based on hind limb development for stages less than 57. Stage 58 is characterized by the forelimbs breaking through the skin.

The second experiment was conducted in a manner similar to the first with a few exceptions. The tadpoles were supplied by Jackson Gross at UW-Madison, with fertilized eggs from four clutches. The concentrations of sodium perchlorate used were 62, 250, 1,000, and 4,000 ppb. The tanks were 10 L with 4-L of test solution. Instead of daily renewal, a flow-through system was used with two turnovers (8 L) of water daily. Feeding was supplemented with Nasco frog brittle (Fort Atkinson, Wis.) at two feedings per day. The experiment was ended at 21 rather than 20 days, and the tadpoles were categorized to the actual NF stage.

## 8.3 Results and Discussion

In the first experiment, the tadpoles in the effluent had begun to complete metamorphosis, but tadpoles growing in the lab water lagged behind (Table 8-1). Tadpoles in one control tank in particular developed more slowly than those in the other treatments. Mortality was very low; only three tadpoles in the entire study died. These results were unexpected in that the researchers did not see any marked differences in metamorphic rates at any of the concentrations tested. Tietge et al. (2005) found delayed metamorphosis in *X. laevis* at concentrations of 250 ppb and above. Additionally, it was unexpected that the metamorphic rates for the tadpoles reared in the secondary treated effluent would be enhanced. Analyses of perchlorate were conducted and were very close to the nominal. One factor is that Tietge et al. are in Duluth, Minn., and the water they use is very soft and has very low iodine levels. Differences such as these may account for differences in the effect of sodium perchlorate.

**Table 8-1. Effects of Perchlorate on Development and Mortality of *X. tropicalis*.**

| | Effluent | 0 ppb | 30 ppb | 62 ppb | 250 ppb | 1,000 ppb |
|---|---|---|---|---|---|---|
| # < Stage 54 | 3 | 9 | 2 | 2 | 5 | 6 |
| # > Stage 58 | 17 | 6 | 6 | 5 | 1 | 3 |
| Mortality (%) | 3 | 6 | 0 | 0 | 0 | 0 |

The results for the second experiment were encouraging in that the control tadpoles developed well, and the researchers had very low mortality. Again, the tadpoles in the secondary treated effluent developed more quickly than in any other treatment (Figures 8-1, 8-2). Comparing the average stages of development (Figure 8-2), ANOVA analysis was significant ($P < 0.001$), with Tukey's pairwise comparison indicating all perchlorate treatments were significantly different than the effluent, but the effluent was not significantly different from the controls ($P = 0.10$). Comparing the perchlorates to the control indicated only a trend towards significance (0 vs. 4,000 ppb $P = 0.07$).

A more sensitive endpoint for *X. laevis* exposure to perchlorate was changes in thyroid histology. Histological changes such as colloid depletion occurred at levels as low as 16 ppb (Tietge et al., 2005). Follicular cell hypertrophy and glandular hypertrophy occurred at 63 ppb (Tietge et al., 2005). Degitz et al. (2005) also found the *X. laevis* model to respond appropriately to two other thyroid system inhibitors (6-propylthiouracil, methimazole) and stimulator (T3).

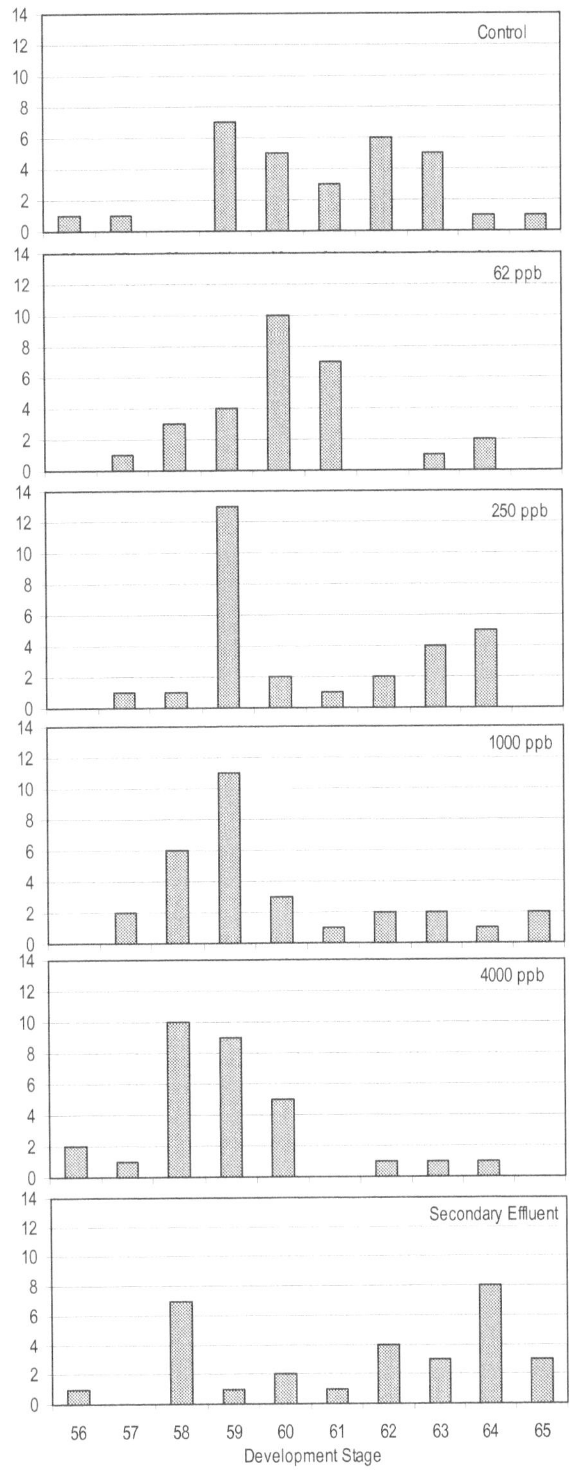


**Figure 8-1.** Number of Tadpoles in Each Developmental Stage After Exposure to Sodium Perchlorate or Secondary Treated Effluent.

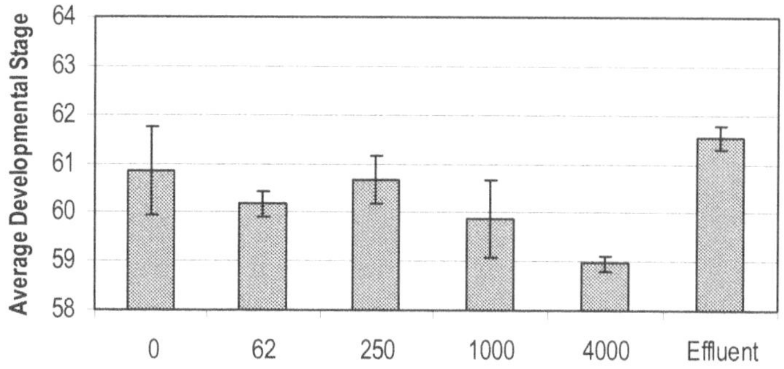


**Figure 8-2. Average Stage for *X. tropicalis* After Exposure to Sodium Perchlorate or Secondary Treated Effluent**. Values represent means ± s.d. n = 3 replicate tanks.

Tietge et al. (2005) found differences in development rates even though they ended the assay after two weeks rather than the three weeks the team used. One reason a shorter duration can be helpful is that variation within a treatment in the stage of development is not as pronounced. The researchers continued the assay for three weeks because no differences were obvious after two weeks of exposure, and the researchers wanted to know if tadpoles at higher perchlorate concentrations could continue metamorphosis. They did find tadpoles at the last two developmental stages (64 and 65) in all treatments, including the highest. Tadpoles are considered frogs when they reach Stage 66.

Given the lack of profound effects in the experiments, the research team did not do histological evaluations of the thyroid, as any changes would not appear to have biological relevance in terms of metamorphic rates. The results emphasize the need for additional endpoints other than developmental stage that are more specific to alterations in the thyroid system. It remains unclear why the researchers did not see as profound an inhibition of metamorphosis as was seen by Tietge et al. (2005) or why developmental rates were enhanced in secondary treated effluents.

## 8.4 Conclusions

Although the researchers were successful in rearing *X. tropicalis,* they were not successful in terms of indicating thyroid disruption. This type of in vivo assay is very time and resource intensive and would be more so if all the appropriate endpoints (histological, thyroid hormone levels) were included.

Overall, the method does show promise, especially if some gene expression biomarkers can be determined. Using gene expression biomarkers could allow for reduced exposure time and likely greater sensitivity.

# APPENDIX A

# SUMMARY OF UTILITY SURVEY FOR FIELD SAMPLING

**Facility A:**

***Treatment***

Facility A is designed to treat 50-mgd and currently operates at 48-mgd. It receives approximately 63% domestic sewage from a metropolitan area and 37% industrial discharge including wastewater from a major food processing company. The service area has no combined sewers, therefore no stormwater runoff is treated. Drinking water supply in the service area of the facility is entirely based on groundwater.

Treatment at Facility A consists of grit removal, primary clarification, activated sludge treatment with biological phosphorus removal and nitrification/denitrification (through a sequence of anaerobic, anoxic, and aerobic zones), secondary clarification, and ultraviolet disinfection prior to discharge. The entire flow is split into half after the grit tanks to serve an east and west treatment train, each operated at 21-mgd. The east and west plants operate three and two, respectively, parallel activated sludge trains followed by secondary clarifiers. A junction structure after secondary treatment combines all flows after secondary clarification prior to UV disinfection. The process schematic of the west plant is illustrated in Figure A-1.

The hydraulic retention time (HRT) during primary treatment is approximately 2.22 hrs at 42 mgd. The average mixed liquor aeration time is 13.7 hours and the solid retention time (SRT) is approximately 10 days. Approximately 93% of the treated effluent is used to augment flow in a Creek and approximately 7% of the total effluent is discharged to another Creek.

The facility employs on-site solids treatment for primary and excess activated sludge. Primary sludge is screened, degritted, and thickened by gravity prior to anaerobic digestion. The supernatant of the gravity sludge thickener (GST) is returned to the primary clarifiers of the East Plant. Digester dewatering is returned to the primary sludge screening. The digested sludge is further thickened by gravity belt thickener (GBT). Excess biosolids are thickened by dissolved air flotation (DAF) prior to anaerobic digestion. The supernatent of DAF and GBT are combined and returned to the grit tanks prior to primary clarification.

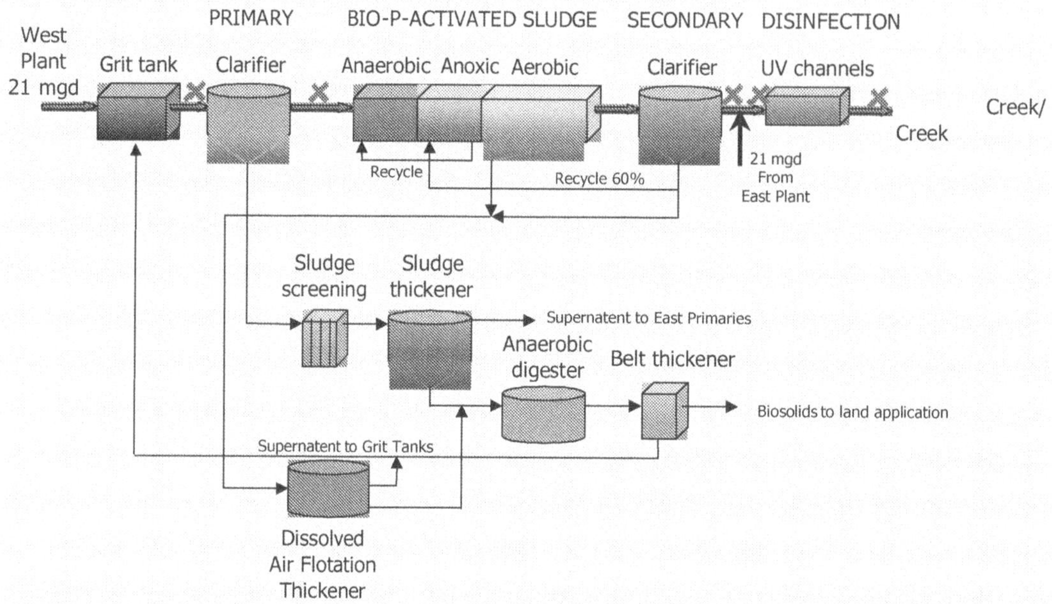


**Figure A-1. Process Schematic of Facility A (✗ Represents Proposed Water Quality Sampling Location).**

### *Water Quality*

Key operational parameters and water quality of secondary treated wastewater at Facility A are summarized in Table A-1 presenting monthly averages values, minima and maxima for the time period 2000-2002.

Table A-1. Key Operational Parameters and Water Quality of Secondary Treated Effluent at Facility A.

| Parameter | Unit | Concentration |
|---|---|---|
| BOD loading | mg/L | 216<br>Min = 91<br>Max = 312 |
| Mixed liquor total suspended solids | mg/L | 1,170<br>Min = 926<br>Max = 3,620 |
| HRT primary | Hrs | 2.22 |
| HRT secondary | Hrs | 13.7 |
| SRT | Days | 10 |
| pH | - | 7.4<br>Min = 7.1<br>Max = 7.8 |
| Conductivity | μS/cm | N/A |
| BOD | mg/L | 3<br>Min = 2<br>Max = 17 |
| COD | mg/L | N/A |
| DOC | mg/L | N/A |
| UV-254 absorbance | 1/m | N/A |
| $NH_3$-N | mg/L | 0.07<br>Min = 0.01<br>Max = 1.02 |
| $NO_3^-$-N | mg/L | 13.2<br>Min = 9.2<br>Max = 15.6 |
| Total P | mg/L | 0.36<br>Min = 0.07<br>Max = 1.49 |

### *Sampling Locations*

Facility A is the only facility in this research study employing biological phosphorus removal with integrated nitrification/denitrification by operating three different redox zones in sequence employing long HRTs and SRTs. Suggested sampling locations for 4-hr composite samples (Table A-2) to examine the removal efficiencies of endocrine disrupting activity were raw sewage at meter vault 2 (split to feed primary clarifiers of West Plant), primary effluent at primary clarifier weir, secondary effluent at secondary clarifier weir, secondary effluent prior to UV disinfection at UV influent outer well (combined secondary effluents from East and West Plants), and UV treated secondary effluent after UV channels. The 4-hr composite samples at each unit operation considered the hydraulic retention time of each unit.

Table A-2. Sampling Locations and Target Parameters for Facility A.

| Sample Location | Type of Sample and Timing | Parameter | Responsible |
|---|---|---|---|
| Raw sewage at meter vault 2 | 4-hr composite Time: 0 | BOD, pH, T, $NH_3$, $NO_3$, total P | Utility |
| | | DOC, UV, conductivity, T, pH, ELISA | CSM |
| | | pH, T; GC-MS; bioassays | WSLH |
| Primary clarifier weir | 4-hr composite Time: 0 + 2.5 hrs | BOD, pH, T | Utility |
| | | DOC, UV, conductivity, T, pH, ELISA | CSM |
| | | pH, T; GC-MS; bioassays | WSLH |
| Secondary clarifier weir | 4-hr composite Time: 0 + 18 hrs | BOD, pH, T, MLTSS, $NH_3$, $NO_3$, total P | Utility |
| | | DOC, UV, conductivity, T, pH, ELISA | CSM |
| | | pH, T; GC-MS; bioassays | WSLH |
| UV influent outer well | 4-hr composite Time: 0 | pH, T | Utility |
| | | DOC, UV, conductivity, T, pH, ELISA | CSM |
| | | pH, T, GC-MS; bioassays | WSLH |
| After UV channel | 4-hr composite Time: 0 + 1 hrs | pH, T | Utility |
| | | DOC, UV, conductivity, T, pH, ELISA | CSM |
| | | pH, T; GC-MS; bioassays | WSLH |

**Facility B:**

***Treatment***

Facility B is designed to treat 28-mgd and currently operates at 23.9-mgd. It receives approximately 95% domestic sewage and 5% industrial discharge. The service area has no combined sewers, therefore no stormwater runoff is treated. Drinking water supply in the service area of the facility is entirely based on surface water.

Treatment at Facility B consists of grit removal, primary clarification, activated sludge treatment with chemical phosphorus removal and partly nitrification (ferric chloride is used for phosphorus precipitation), secondary clarification, disinfection using chlorine, and dechlorination using sulfur dioxide prior to discharge to a lake. The process schematic of Facility B is illustrated in Figure A-2.

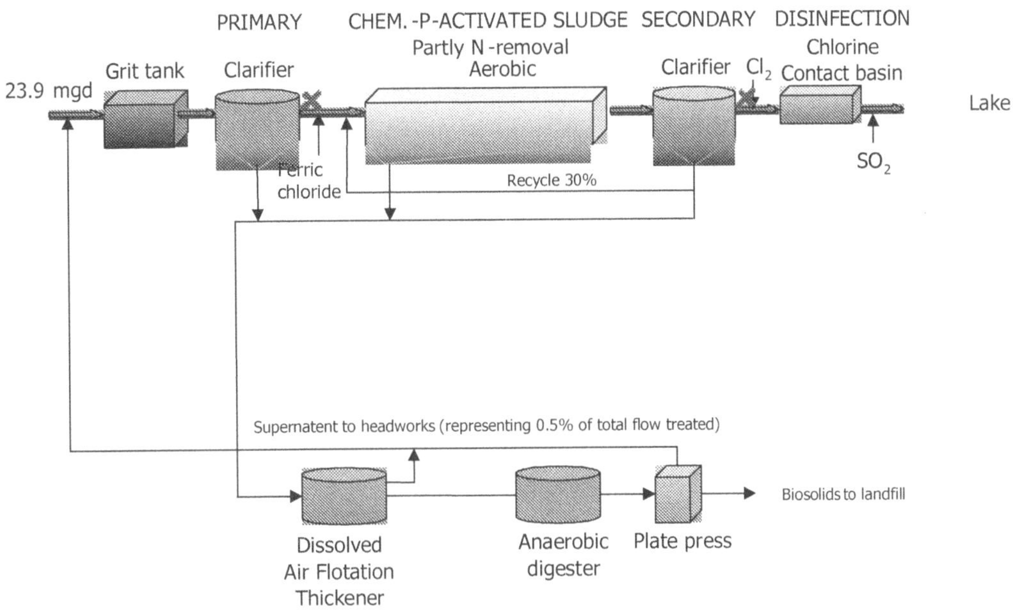


**Figure A-2. Process Schematic of Facility B (✖ Represents Water Quality Sampling Location).**

The HRT during primary treatment is approximately 2.9 hrs. The average mixed liquor aeration time is 5.5 hours for the current average flow and the solid retention time is approximately 8.5 days. During average flow conditions, the HRT in the chlorine contact basins is 74 min. Approximately 99% of the treated effluent is discharged to the lake, less than 1% of the flow is used onsite for make-up water for thickener units and chlorination/dechlorination systems.

The facility employs on-site solids treatment for primary and excess activated sludge. Primary and excess biosolids are thickened by dissolved air flotation (DAF) prior to anaerobic digestion. The digested biosolids are dewatered using a sludge press and hauled to a landfill. The supernatant of DAF and the sludge press is returned to the headworks representing 0.5% of the total water flow treated.

### *Water Quality*

Key operational parameters and water quality of secondary treated wastewater at Facility B are summarized in Table A-3 presenting monthly average values, minima and maxima for the time period 1999-2001.

Table A-3. Key Operational Parameters and Water Quality of Secondary Treated Effluent at Facility B.

| Parameter | Unit | Concentration |
|---|---|---|
| BOD loading | mg/L | 140<br>Min = 134<br>Max = 145 |
| Mixed liquor total suspended solids | mg/L | 2400<br>Min = 2304<br>Max = 2509 |
| HRT primary | Hrs | 2.9 |
| HRT secondary | Hrs | 5.5 |
| SRT | Days | 8.5 |
| pH | - | 7.3<br>Min = 7.1<br>Max = 7.5 |
| Conductivity | µS/cm | N/A |
| BOD | mg/L | 10<br>Min = 3<br>Max = 52 |
| COD | mg/L | N/A |
| DOC | mg/L | N/A |
| UV-254 absorbance | 1/m | N/A |
| $NH_3$-N | mg/L | 7.2<br>Min = 0.5<br>Max = 17 |
| $NO_3^-$-N | mg/L | N/A |
| $PO_4^{3-}$-P | mg/L | 0.5<br>Min = 0.22<br>Max = 0.84 |

### *Sampling Locations*

For background comparison, a surface water sample was collected from the lake upstream of the discharge point (Table A-4).

Table A-4. Sampling Locations and Target Parameters for Facility B.

| Sample Location | Type of Sample and Timing | Parameter | Responsible |
|---|---|---|---|
| Lake | Grab sample | pH, T, dissolved oxygen | Utility, WSLH |
| | | DOC, UV, conductivity, T, pH, ELISA | CSM |
| | | pH, T; GC-MS; bioassays | WSLH |
| Primary clarifier weir | 4-hr composite<br>Time: 0 | BOD, pH, T, $NH_3$-N, $PO_4$-P | Utility |
| | | DOC, UV, conductivity, T, pH, ELISA | CSM |
| | | pH, T, $NO_3^-$-N GC-MS; bioassays | WSLH |
| Secondary clarifier weir | 4-hr composite<br>Time: 0 + 10 hrs | BOD, pH, T, MLTSS, $NH_3$, $PO_4$-P | Utility |
| | | DOC, UV, conductivity, T, pH, ELISA | CSM |
| | | pH, T, $NO_3^-$-N; GC-MS; bioassays | WSLH |

Two composite samples (Table A-4) were collected at the WWTP to examine the removal efficiencies of endocrine disrupting activity during activated sludge treatment with chemical P-removal. A primary effluent composite sample at the weir of the primary clarifier and a secondary treated effluent prior to chlorination were collected. The 4-hr composite samples at each unit operation considered the hydraulic retention time of each unit.

**Facility C:**
***Treatment***

Facility C is designed to treat 36-mgd and currently operates at design capacity. It receives approximately 95% domestic sewage and 5% industrial discharge. The service area has no combined sewer systems, therefore no stormwater runoff is treated. Drinking water supply in the service area of the facility is primarily based on surface water and very little groundwater.

Treatment at Facility C consists of grit removal, pre-aeration, primary clarification, activated sludge treatment (not nitrifying), secondary clarification, and disinfection using sodium hypochlorite. A small portion of the treated effluent is used onsite for landscape irrigation. The majority of the treated effluent is discharged to the Ocean approximately 1.5 miles off shore. The process schematic of Facility C is illustrated in Figure A-3. The hydraulic retention time during primary treatment is approximately 3.8 hrs at 36 mgd. The average mixed liquor aeration time is 1.2 hours and the solid retention time is approximately 1.7 days.

The facility employs on-site solids treatment for primary and excess activated sludge. On-site biosolids treatment consists of anaerobic mesophilic digestion for 15-30 days followed by centrifuge dewatering, approximately 30-90 days storage on a covered open air pad and finally is land applied as a soil conditioner. Digesters are not supernated. The supernatent of the centrifuge dewatering is returned to the headworks.

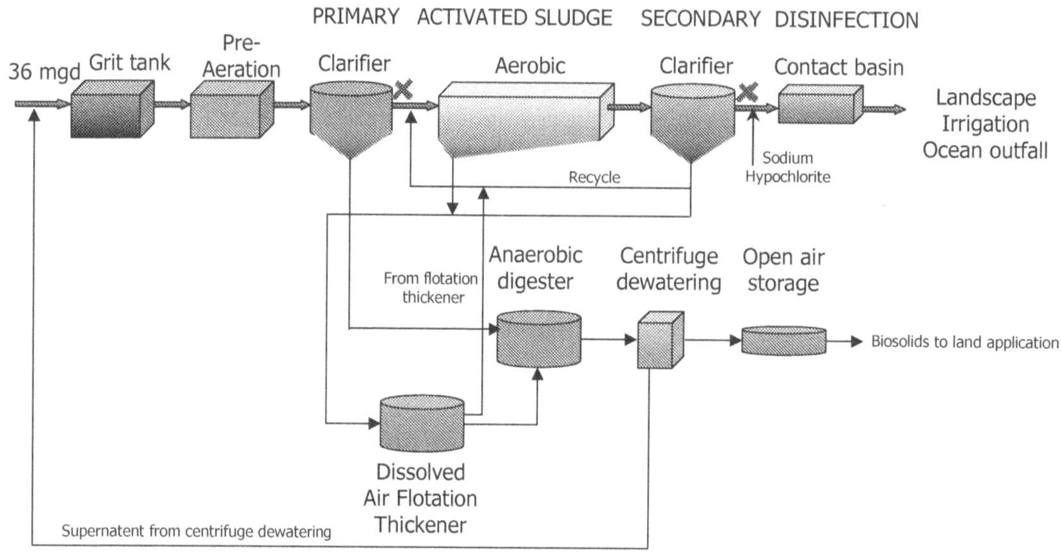


**Figure A-3. Process Schematic of Facility C (✖ Represents Water Quality Sampling Location).**

### *Water Quality*

Key operational parameters and water quality of secondary treated wastewater at Facility C are summarized in Table A-5 presenting monthly average values, minima and maxima for the time period 1999-2002.

**Table A-5. Key Operational Parameters and Water Quality of Secondary Treated Effluent at Facility C.**

| Parameter | Unit | Concentration |
|---|---|---|
| BOD loading | kg/day | 24,716<br>Min = N/A<br>Max = N/A |
| Mixed liquor total suspended solids | mg/L | 2,685<br>Min = N/A<br>Max = N/A |
| HRT primary | Hrs | 3.8 |
| HRT secondary | Hrs | 1.2 |
| SRT | Days | 1.7 |
| PH | - | 6.6<br>Min = 6.3<br>Max = 6.8 |
| Conductivity | μS/cm | N/A |
| BOD | mg/L | 11<br>Min = 4<br>Max = 18 |
| COD | mg/L | N/A |
| DOC | mg/L | N/A |
| UV-254 absorbance | 1/m | N/A |
| $NH_3$-N | mg/L | N/A |
| $NO_3^-$-N | mg/L | N/A |
| $PO_4^{3-}$-P | mg/L | N/A |

### *Sampling Locations*

For background comparison, a surface water sample was collected from the surface water source (Table A-6). Two composite samples were collected at Facility C to examine the removal efficiencies of endocrine disrupting activity during activated sludge treatment employing very short HRTs and no nitrification. A primary effluent sample at the weir of the primary clarifier and a secondary treated effluent prior to chlorination were collected. The 4-hr composite samples at each unit operation considered the hydraulic retention time of each unit.

**Table A-6. Sampling Locations and Target Parameters for Facility C.**

| Sample Location | Type of Sample and Timing | Parameter | Responsible |
|---|---|---|---|
| Surface Water | Grab sample | pH, T, dissolved oxygen | Utility, WSLH |
| | | DOC, UV, conductivity, T, pH, ELISA | CSM |
| | | pH, T<br>GC-MS; bioassays | WSLH |
| Primary clarifier weir | 4-hr composite<br>Time: 0 | pH, T, conductivity, $NH_3$-N, $NO_3$-N, $PO_4$-P | Utility |
| | | DOC, UV, conductivity, T, pH, ELISA | CSM |
| | | pH, T<br>GC-MS; bioassays | WSLH |
| Secondary clarifier weir | 4-hr composite<br>Time: 0 + 3 hrs | pH, T, MLTSS, conductivity, $NH_3$, $NO_3$-N, $PO_4$-P | Utility |
| | | DOC, UV, conductivity, T, pH, ELISA | CSM |
| | | pH, T<br>GC-MS; bioassays | WSLH |

## Facility D:

### *Treatment*

Facility D is designed to treat 62.5-mgd and currently operates at design capacity. It receives approximately 90% residential sewage and 10% industrial discharge. The service area has no combined sewers, therefore no stormwater runoff is treated. Drinking water supply in the service area of the facility is a blend of surface and groundwater.

Treatment at Facility D consists of primary clarification, activated sludge treatment with partly nitrification, secondary clarification, alum addition for coagulation prior to filtration, disinfection using chlorine gas (pre-filter chlorination is used), tertiary dual-media filtration (anthracite, sand), and chlorine contact basin. Sulfur dioxide is used after chlorine contact to dechlorinate prior to discharge. The process schematic of Facility D is illustrated in Figure A-4.

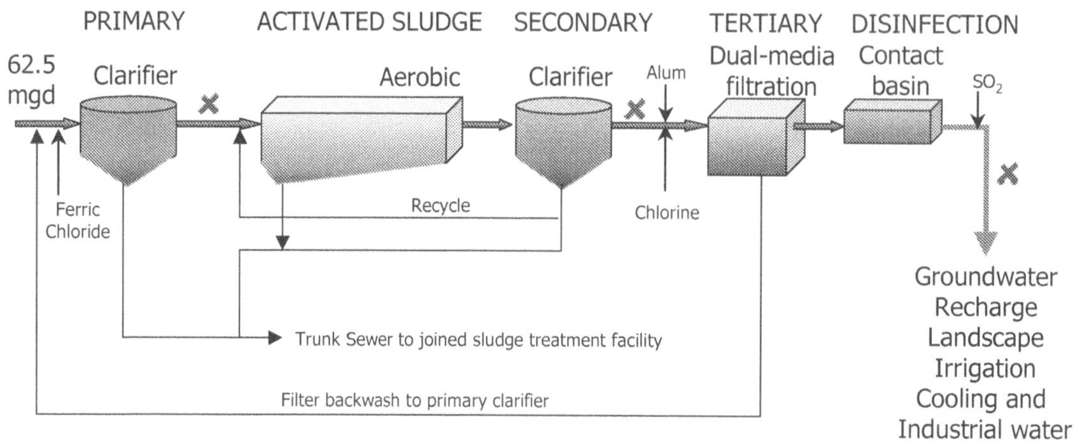


**Figure A-4. Process Schematic of Facility D ( ✗ Represents Water Quality Sampling Locations).**

The hydraulic retention time during primary treatment is approximately 1.6 hrs at 62.5-mgd. The average mixed liquor aeration time is 2.3 hours and the solid retention time is approximately 10 days. The average detention time in the dual-media filters is 17 minutes. The contact time in the chlorination contact basin is approximately 90 minutes. For fiscal year 2001-2002, 54.3% of the effluent produced at Facility D was actively reused for groundwater recharge and irrigation purposes. The great majority (92.5%) of the reclaimed water was used to recharge the groundwater aquifer and 5.7% of reclaimed water was used for irrigation of landscape, agriculture, ornamental plants, and athletic fields. The remaining 1.8% was reused for cooling and industrial processes at a local landfill.

The facility does not employ on-site solids treatment. Excess sludge is discharged and transported via sewer to a centralized treatment facility.

### *Water Quality*

Key operational parameters and water quality of secondary treated wastewater at Facility D are summarized in Table A-7 presenting monthly average values, minima and maxima for the time period 1997-2002. The effluent quality meets California Title 22 standards for recycled water.

Table A-7. Key Operational Parameters and Water Quality of Secondary Treated Effluent at Facility D.

| Parameter | Unit | Concentration |
|---|---|---|
| BOD loading | lbs/day | 88,000<br>Min = 57,000<br>Max = 99,600 |
| Mixed liquor total suspended solids | mg/L | 1,550<br>Min = 1,260<br>Max = 1,640 |
| HRT primary | Hrs | 1.6 |
| HRT secondary | Hrs | 2.3 |
| SRT | Days | 10 |
| pH | - | 6.9<br>Min = 6.8<br>Max = 7.2 |
| Conductivity | µS/cm | 1,150<br>Min = 1,050<br>Max = 1,260 |
| BOD | mg/L | 4<br>Min = 2<br>Max = 10.2 |
| COD | mg/L | 27<br>Min = 23<br>Max = 38 |
| DOC | mg/L | 9<br>Min = 5.1<br>Max = 11 |
| UV-254 absorbance | 1/m | N/A |
| $NH_3$-N | mg/L | 9<br>Min = 5.3<br>Max = 20.2 |
| $NO_3^-$-N | mg/L | 3.4<br>Min = 0.2<br>Max = 4.4 |
| $PO_4^{3-}$-P | mg/L | 2.3<br>Min = 1.6<br>Max = 4.4 |

### *Sampling Locations*

Three composite samples were collected at Facility D to examine the removal efficiencies of endocrine disrupting activity during secondary and tertiary (with alum addition) treatment (Table A-8). A primary effluent sample at the weir of the primary clarifier, a secondary treated effluent at the weir of the secondary clarifier prior to chlorination and alum addition, and a final effluent sample after dechlorination were collected. The 4-hr composite samples at each unit operation considered the hydraulic retention time of each unit.

Table A-8. Sampling Locations and Target Parameters for Facility D.

| Sample Location | Type of Sample and Timing | Parameter | Responsible |
|---|---|---|---|
| Primary effluent at weir of primary clarifier | 4-hr composite Time: 0 | pH, T, conductivity, $NH_3$-N, $NO_3$-N, $PO_4$-P | Utility |
| | | DOC, UV, conductivity, T, pH, ELISA | CSM |
| | | pH, T GC-MS; bioassays | WSLH |
| Secondary effluent at weir of secondary clarifier | 4-hr composite Time: 0 + 2.5 hrs | pH, T, MLTSS, conductivity, $NH_3$, $NO_3$-N, $PO_4$-P | Utility |
| | | DOC, UV, conductivity, T, pH, ELISA | CSM |
| | | pH, T GC-MS; bioassays | WSLH |
| Tertiary effluent after dechlorination | 4-hr composite Time: 0 + 4.5 hrs | pH, T, conductivity, $NH_3$, $NO_3$-N, $PO_4$-P | Utility |
| | | DOC, UV, conductivity, T, pH, ELISA | CSM |
| | | pH, T GC-MS; bioassays | WSLH |

## Facility E:

### *Treatment*

Facility E is designed to treat 18-mgd and currently operates at 9.1-mgd. It receives approximately 90% residential sewage and 10% industrial discharge primarily from semi-conductor industry. The service area has no combined sewers, therefore no stormwater runoff is treated. Drinking water supply in the service area of the facility is a blend of surface and groundwater. The groundwater originates from the local aquifer and represents approximately 7% of the drinking water supply.

Treatment at Facility E consists of grit removal, primary clarification, activated sludge treatment with nitrification/denitrification through a sequence of anoxic and aerobic zones, secondary clarification, tertiary dual-media filtration (anthracite, sand), and UV disinfection. The process schematic of Facility E is illustrated in Figure A-5.

The hydraulic retention time during primary treatment is approximately 2.3 hrs at 9.1-mgd. The average mixed liquor aeration time is 5 hours and the solid retention time is approximately 10 days. The average detention time in the dual-media filters is 20 minutes. Approximately 90% of the tertiary effluent is used for groundwater recharge using surface spreading basins. The remaining 10% are discharged to a river after UV disinfection.

The facility employs on-site solids treatment for primary and excess activated sludge. On-site biosolid treatment consists of sludge thickening using centrifuges and mesophilic anaerobic digestion. The supernatent of the digesters and the centrate are returned to the headworks.

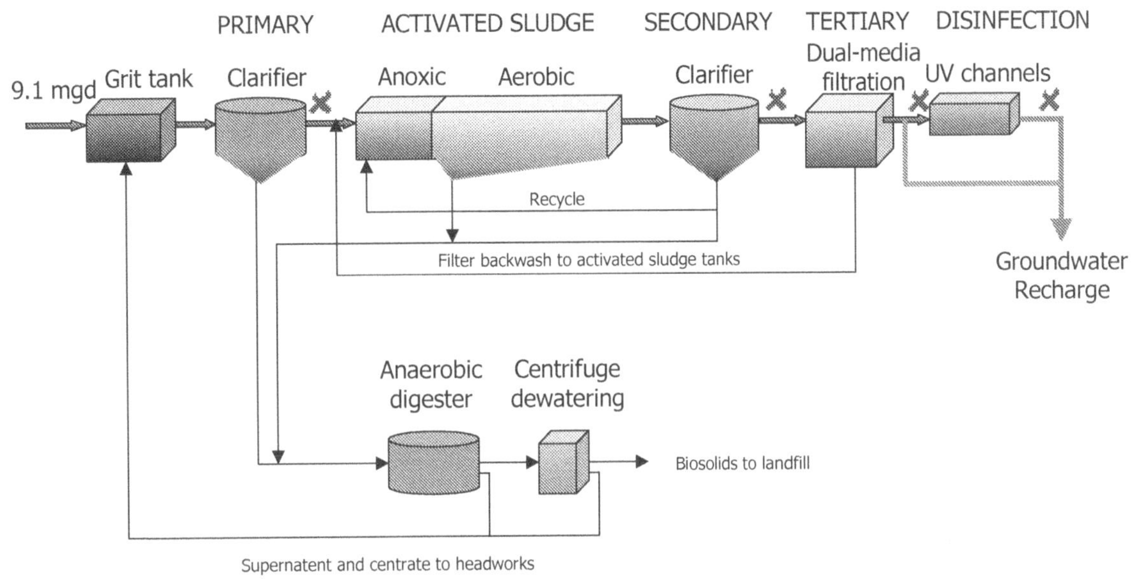


**Figure A-5. Process Schematic of Facility E ( ✖ Represents Water Quality Sampling Locations).**

### *Water Quality*

Key operational parameters and water quality of secondary treated wastewater at Facility E are summarized in Table A-9 presenting monthly average values, minima and maxima for the time period 1997-2002.

**Table A-9. Key Operational Parameters and Water Quality of Secondary Treated Effluent at Facility E.**

| Parameter | Unit | Concentration |
|---|---|---|
| BOD loading | mg/L | 186<br>Min = N/A<br>Max = 215 |
| Mixed liquor total suspended solids | mg/L | 3,000<br>Min = N/A<br>Max = N/A |
| HRT primary | Hrs | 2.3 |
| HRT secondary | Hrs | 5 |
| SRT | Days | 10 |
| pH | - | 7.5<br>Min = N/A<br>Max = N/A |
| Conductivity | μS/cm | 1,580<br>Min = N/A<br>Max = N/A |
| BOD | mg/L | N/A |
| COD | mg/L | N/A |
| DOC | mg/L | 5.7<br>Min = 3.9<br>Max = 6.6 |
| UV-254 absorbance | 1/m | 7.0<br>Min = 4.6<br>Max = 9 |
| $NH_3$-N | mg/L | < 0.1<br>Min = N/A<br>Max = N/A |
| $NO_3^-$-N | mg/L | 2.2<br>Min = 0.84<br>Max = 7.1 |
| $PO_4^{3-}$-P | mg/L | 11.4<br>Min = 6.89<br>Max = 10.6 |

### *Sampling Locations*

Four composite samples were collected at Facility E to examine the removal efficiencies of endocrine disrupting activity during secondary treatment (nitrification/denitrification), tertiary filtration, and UV disinfection (Table A-10): a primary effluent sample at the weir of the primary clarifier, a secondary treated effluent at the weir of the secondary clarifier, a tertiary effluent sample at the influent of the UV channel, and a final effluent sample after UV treatment. The 4-hr composite samples at each unit operation considered the hydraulic retention time of each unit.

Table A-10. Sampling Locations and Target Parameters for Facility E.

| Sample Location | Type of Sample and Timing | Parameter | Responsible |
|---|---|---|---|
| Primary effluent at weir of primary clarifier | 4-hrs composite Time: 0 hrs | pH, T, conductivity, $NH_3$-N, $NO_3$-N, $PO_4$-P | Utility |
| | | DOC, UV, conductivity, T, pH, ELISA | CSM |
| | | pH, T GC-MS; bioassays | WSLH |
| Secondary effluent at weir of secondary clarifier | 4-hr composite Time: 0 + 7 hrs | pH, T, MLTSS, conductivity, $NH_3$, $NO_3$-N, $PO_4$-P | Utility |
| | | DOC, UV, conductivity, T, pH, ELISA | CSM |
| | | pH, T GC-MS; bioassays | WSLH |
| Tertiary effluent after dual media filtration at influent of UV channel | 4-hr composite Time: 0 + 8 hrs | pH, T, conductivity, $NH_3$, $NO_3$-N, $PO_4$-P | Utility |
| | | DOC, UV, conductivity, T, pH, ELISA | CSM |
| | | pH, T GC-MS; bioassays | WSLH |
| Final effluent after UV treatment at effluent of UV channel | 4-hr composite Time: 0 + 8.5 hrs | pH, T, conductivity, $NH_3$, $NO_3$-N, $PO_4$-P | Utility |
| | | DOC, UV, conductivity, T, pH, ELISA | CSM |
| | | pH, T GC-MS; bioassays | WSLH |

## Facility F:

### *Treatment*

This facility is permitted as a 30.5-mgd design capacity facility and consists of three plants: north, southeast, and southwest. The facility is currently expanded to 50-mgd. The effluent of the three plants can be directed to separate pump stations and they may be blended together. The north plant is selected as Facility F for this study. The north Plant is designed to treat 23-mgd and currently operates at 17-mgd. The service area has combined sewers, but stormwater runoff occurs only seasonally with an average of less than 1-mgd per month. Drinking water supply in the service area of the facility is based on local groundwater.

Treatment at Facility F consists of grit removal, activated sludge treatment, secondary clarification, flow equalization basin, flocculation, tertiary dual-media filtration (anthracite, sand), and high-level disinfection using chlorine. The process schematic of Facility F is illustrated in Figure A-6. The average mixed liquor aeration time is 6.8 hours at 17-mgd and the solid retention time is approximately 4.4 days. The average detention time in the dual-media filters is 27 minutes. The contact time in the chlorination contact basin is approximately 30 minutes. The reclaimed water from the facility is reclaimed for irrigation, rapid infiltration basins, golf course irrigation, small commercial and residential irrigation, and other uses.

The facility employs on-site solids treatment for primary and excess activated sludge. On-site biosolid treatment consists of sludge thickening using dissolved air flotation, anaerobic digestion followed by dewatering using belt filter press units prior to land application. The supernatent of the digesters is returned to the headworks.

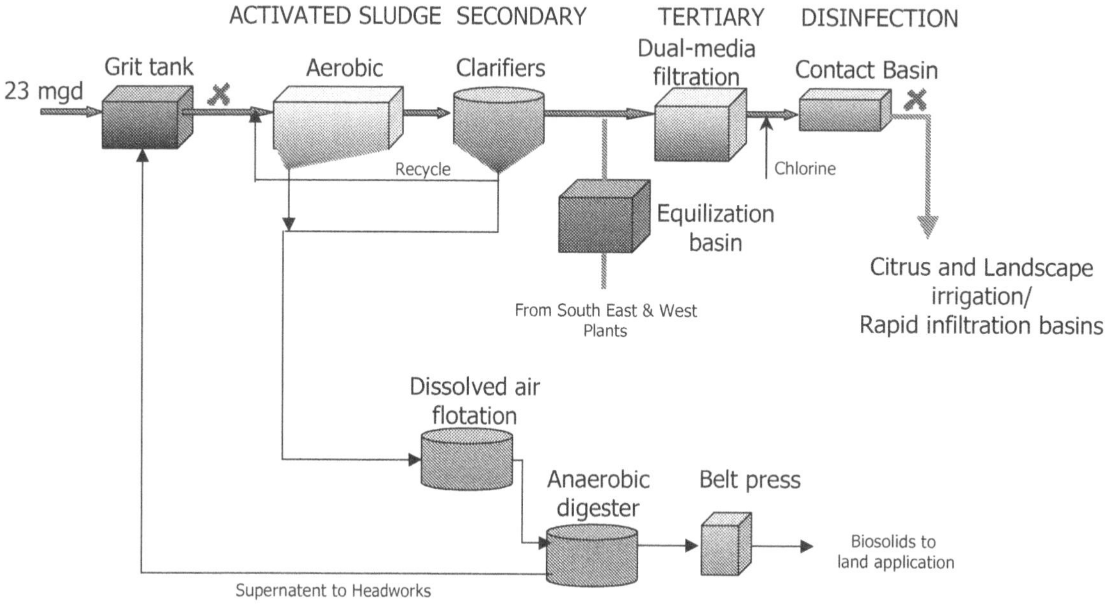


**Figure A-6. Process Schematic of Facility F ( ✖ Represents Water Quality Sampling Locations).**

### *Water Quality*

Key operational parameters and water quality of secondary treated wastewater at Facility F are summarized in Table A-11 presenting monthly average values, minima and maxima for the time period 1997-2001.



WERF

Table A-11. Key Operational Parameters and Water Quality of Secondary Treated Effluent at Facility F.

| Parameter | Unit | Concentration |
|---|---|---|
| BOD loading | mg/L | 150<br>Min = 95<br>Max = 210 |
| Mixed liquor total suspended solids | mg/L | 3,500<br>Min = 2,705<br>Max = 4,516 |
| HRT secondary | Hrs | 6.8 |
| SRT | Days | 4.4 |
| pH | - | 7.4<br>Min = 6.9<br>Max = 8.5 |
| Conductivity | μS/cm | 650<br>Min = 601<br>Max = 745 |
| BOD | mg/L | 1.8<br>Min = 1.0<br>Max = 3.0 |
| COD | mg/L | N/A |
| DOC | mg/L | N/A |
| UV-254 absorbance | 1/m | N/A |
| $NH_3$ | mg/L | 2.5<br>Min = 1.19<br>Max = 5.5 |
| $NO_3^-$-N | mg/L | 5.88<br>Min = 3.05<br>Max = 7.79 |
| $PO_4^{3-}$-P | mg/L | 1.9<br>Min = 1.23<br>Max = 3.13 |

### *Sampling Locations*

Two composite samples were collected at Facility F to examine the removal efficiencies of endocrine disrupting activity during activated sludge treatment, tertiary filtration, and chlorination (Table A-12). Two two-hour composite samples were collected at the influent of the aeration tanks and after the chlorine contact chamber. The 4-hr composite sample after chlorination considered the hydraulic retention time of all unit processes.

Table A-12. Sampling Locations and Target Parameters for Facility F.

| Sample Location | Type of Sample and Timing | Parameter | Responsible |
|---|---|---|---|
| Raw effluent prior to aeration basin | 4-hr composite Time: 0 | pH, T, conductivity, $NH_3$-N, $NO_3$-N, $PO_4$-P | Utility |
| | | DOC, UV, conductivity, T, pH, ELISA | CSM |
| | | pH, T, GC-MS; bioassays | WSLH |
| Tertiary effluent after chlorine contact chamber | 4-hr composite Time: 0 + 10 hrs | pH, T, MLTSS, conductivity, $NH_3$, $NO_3$-N, $PO_4$-P | Utility |
| | | DOC, UV, conductivity, T, pH, ELISA | CSM |
| | | pH, T, GC-MS; bioassays | WSLH |

## Facility G:

### *Treatment*

Facility G is designed to treat 63-mgd and currently operates at design capacity. The exact ratio of domestic to industrial sewage is not known. The service area has no combined sewers, therefore no stormwater runoff is treated. Drinking water supply in the service area of the facility is a blend of surface and groundwater. The groundwater originates from the local aquifer. The majority of the drinking water supply is surface water.

Treatment at Facility G consists of grit removal, primary clarification, activated sludge treatment with nitrification/denitrification through a sequence of anoxic and aerobic zones, secondary clarification, tertiary dual-media filtration, and disinfection using chlorine. The process schematic of Facility G is illustrated in Figure A-7. The hydraulic retention time during primary treatment is approximately 2 hours at 63-mgd. The average mixed liquor aeration time is 7.3 hours and the solid retention time is approximately 7 days. The average detention time in the filters is 4.5 hours. The majority of the treated effluent is discharged to a river and diverted for agricultural irrigation.

The facility employs on-site solids treatment for primary and excess activated sludge. On-site biosolid treatment consists of thickening centrifuges for primary sludge and dissolved air flotation for biosolids prior to mesophilic anaerobic digestion with centrifuge dewatering. The centrates are returned to the headworks.

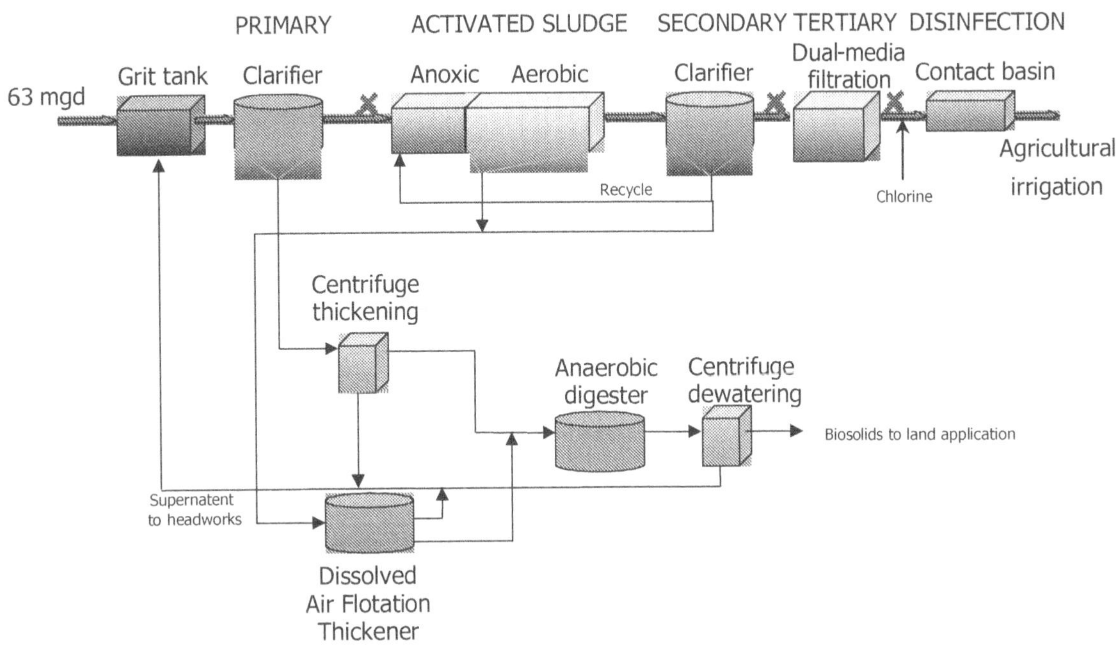


**Figure A-7. Process Schematic of Facility G (✖ Represents Water Quality Sampling Locations).**

### *Water Quality*

Key operational parameters and water quality of secondary treated wastewater at Facility G are summarized in Table A-13 presenting monthly average values, minima and maxima for the time period 1997-2002.

**Table A-13. Key Operational Parameters and Water Quality of Secondary Treated Effluent at Facility G.**

| Parameter | Unit | Concentration |
|---|---|---|
| BOD loading | mg/L | 260<br>Min = 248<br>Max = 285 |
| Mixed liquor total suspended solids | mg/L | 3,000<br>Min = 2,685<br>Max = 3,046 |
| HRT primary | Hrs | 2.0 |
| HRT secondary | Hrs | 7.3 |
| SRT | Days | 7 |
| pH | - | 7.0<br>Min = 6.91<br>Max = 7.10 |
| Conductivity | μS/cm | 1,600<br>Min = 1,450<br>Max = 1,720 |
| BOD | mg/L | 2.18<br>Min = 2.02<br>Max = 2.25 |
| COD | mg/L | N/A |
| DOC | mg/L | 6.75<br>Min = 6.61<br>Max = 7.17 |
| UV-254 absorbance | 1/m | N/A |
| $NH_3$-N | mg/L | 0.46<br>Min = 0.38<br>Max = 0.54 |
| $NO_3^-$-N | mg/L | 5.26<br>Min = 4.84<br>Max = 6.18 |
| $PO_4^{3-}$-P | mg/L | N/A |

### *Sampling Locations*

Three composite samples were collected at Facility G to examine the removal efficiencies of endocrine disrupting activity during secondary treatment (nitrification/denitrification) followed by tertiary filtration (Table A-14). A primary effluent sample at the weir of the primary clarifier, a secondary treated effluent at the weir of the secondary clarifier, and a tertiary treated effluent prior to chlorination were sampled. The 4-hr composite samples at each unit operation considered the hydraulic retention time of each unit.

Table A-14 summarizes the sampling locations and unit operations of investigation.

**Table A-14. Sampling Locations and Target Parameters for Facility G.**

| Sample Location | Type of Sample and Timing | Parameter | Responsible |
|---|---|---|---|
| Primary effluent at weir of primary clarifier | 4-hr composite Time: 0 hrs | pH, T, conductivity, $NH_3$-N, $NO_3$-N, $PO_4$-P | ASU |
| | | DOC, UV, conductivity, T, pH, ELISA | CSM |
| | | pH, T GC-MS; bioassays | WSLH |
| Secondary effluent at weir of secondary clarifier | 4-hr composite Time: 0 + 9 hrs | pH, T, conductivity, $NH_3$, $NO_3$-N, $PO_4$-P | ASU |
| | | DOC, UV, conductivity, T, pH, ELISA | CSM |
| | | pH, T GC-MS; bioassays | WSLH |
| Tertiary effluent after filter prior to chlorine addition | 4-hr composite Time: 0 + 14 hrs | pH, T, conductivity, $NH_3$, $NO_3$-N, $PO_4$-P | ASU |
| | | DOC, UV, conductivity, T, pH, ELISA | CSM |
| | | pH, T GC-MS; bioassays | WSLH |

## Facility H:

### *Treatment*

Facility H is designed to treat 167-mgd and currently operates at 127-mgd. It receives approximately 70% residential sewage, 20% commercial/institutional sewage, and 10% industrial discharge. A majority of the industrial discharge comes from the semiconductor and metal finishing industries. The service area has no combined sewers, therefore no stormwater runoff is treated. Drinking water supply in the service area of the facility is based on surface water and local groundwater provided by the Water District. Surface water used for drinking is imported from other sources.

Treatment at Facility H consists of grit removal, primary clarification, two parallel activated sludge treatment trains (both nitrifying), secondary clarification, tertiary dual-media filtration (anthracite, sand), and disinfection using chlorine, occasionally sodium hypochlorite. Ammonia is added prior to chlorination to achieve chloramination. Both, pre-filter and post-filter chlorinations are used. The process schematic of Facility H is illustrated in Figure A-8. The hydraulic retention time during primary treatment is approximately 1.5 hrs at 127-mgd. The average mixed liquor aeration time is 4-5 hours and the solid retention time is approximately 7 to 8 days. The average detention time in the dual-media filters is 5.2 minutes. The contact time in the chlorination contact basin is approximately 90 minutes. The plant discharges to an area used by a water recycling system. The recycling program delivers recycled water for cooling/industrial processes and landscape irrigation to more than 350 customers. The combined use of the recycling system reached 10 mgd in 2002. The current recycling customer base is primarily landscape irrigation, including parks, golf courses, and cemeteries.

The facility employs on-site solids treatment for primary and excess activated sludge. On-site biosolids treatment consists of sludge thickening, anaerobic digestion followed by storage lagoons and drying beds. The supernatent of the digesters is returned to the headworks.

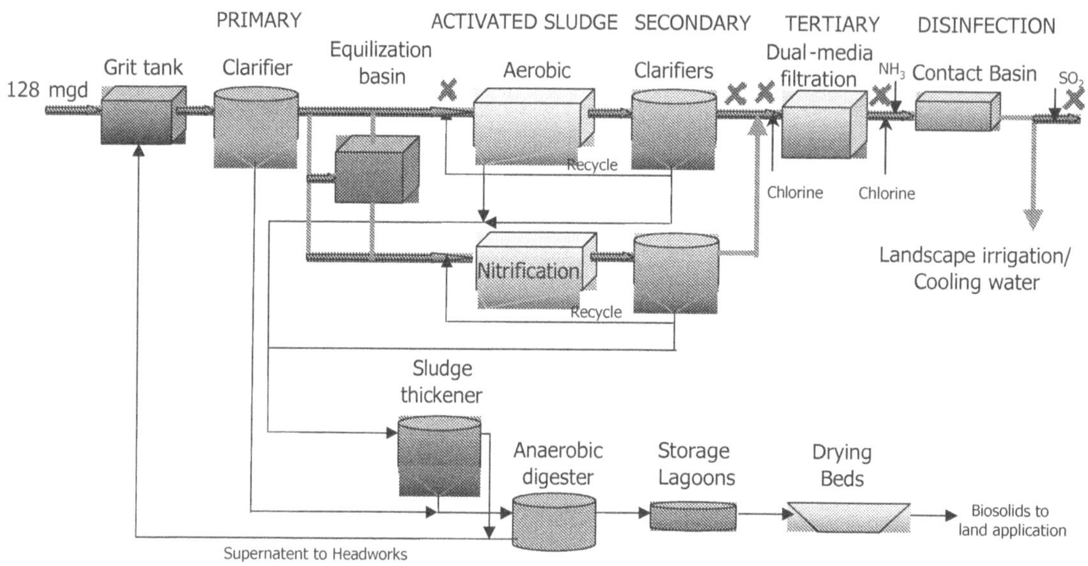


**Figure A-8. Process Schematic of Facility H ( ✗Represents Water Quality Sampling Locations).**

### *Water Quality*

Key operational parameters and water quality of secondary treated wastewater at Facility H are summarized in Table A-15 presenting monthly average values, minima and maxima for the time period 1997-2002. The effluent quality meets California Title 22 standards for recycled water.

Table A-15. Key Operational Parameters and Water Quality of Secondary Treated Effluent at Facility H.

| Parameter | Unit | Concentration |
|---|---|---|
| BOD loading | mg/L | 165<br>Min = 160<br>Max = 166 |
| Mixed liquor total suspended solids | mg/L | 3000<br>Min = N/A<br>Max = N/A |
| HRT primary | Hrs | 1.5 |
| HRT secondary | Hrs | 4-5 |
| SRT | Days | 7-8 |
| pH | - | 7.3<br>Min = N/A<br>Max = N/A |
| Conductivity | μS/cm | 1250<br>Min = N/A<br>Max = N/A |
| BOD | mg/L | 4<br>Min = N/A<br>Max = N/A |
| COD | mg/L | N/A |
| DOC | mg/L | 9<br>Min = N/A<br>Max = N/A |
| UV-254 absorbance | 1/m | N/A |
| $NH_3$-N | mg/L | 0.5<br>Min = N/A<br>Max = N/A |
| $NO_3^-$-N | mg/L | 10.0<br>Min = N/A<br>Max = N/A |
| $PO_4^{3-}$-P | mg/L | 1.3<br>Min = N/A<br>Max = N/A |

***Sampling Locations***

For background comparison, a raw drinking water sample were collected from the surface water source (Table A-16). Five composite samples were collected at Facility H to examine the removal efficiencies of endocrine disrupting activity during activated sludge treatment (nitrification), tertiary filtration, and chlorination. Two 4-hr composite samples were collected prior to the nitrification aeration basin and at the weir of the secondary clarifier. A combined secondary treated effluent composite sample were collected prior to and after tertiary filtration as well as after the chlorine contact basin. The 4-hr composite samples at each unit operation considered the hydraulic retention time of each unit.

Table A-16. Sampling Locations and Target Parameters for Facility H.

| Sample Location | Type of Sample and Timing | Parameter | Responsible |
|---|---|---|---|
| Surface Water | Grab sample | pH, T, dissolved oxygen | Utility, WSLH |
| | | DOC, UV, conductivity, T, pH, ELISA | CSM |
| | | pH, T, GC-MS; bioassays | WSLH |
| Primary effluent prior to nitrification aeration basin | 4-hr composite Time: 0 | pH, T, conductivity, $NH_3$-N, $NO_3$-N, $PO_4$-P | Utility |
| | | DOC, UV, conductivity, T, pH, ELISA | CSM |
| | | pH, T, GC-MS; bioassays | WSLH |
| Secondary effluent at weir of nitrification clarifier | 4-hr composite Time: 0 + 10 hrs | pH, T, MLTSS, conductivity, $NH_3$, $NO_3$-N, $PO_4$-P | Utility |
| | | DOC, UV, conductivity, T, pH, ELISA | CSM |
| | | pH, T, GC-MS; bioassays | WSLH |
| Combined secondary effluent prior to tertiary filtration | 1-hr composite Time: 0 | pH, T, MLTSS, conductivity, $NH_3$, $NO_3$-N, $PO_4$-P | Utility |
| | | DOC, UV, conductivity, T, pH, ELISA | CSM |
| | | pH, T, GC-MS; bioassays | WSLH |
| After tertiary filtration prior to chlorine addition | 1-hr composite Time: 0 + 15 min | pH, T, MLTSS, conductivity, $NH_3$, $NO_3$-N, $PO_4$-P | Utility |
| | | DOC, UV, conductivity, T, pH, ELISA | CSM |
| | | pH, T, GC-MS; bioassays | WSLH |
| After chlorine contact basin | 1-hr composite Time: 0 + 105 min | pH, T, MLTSS, conductivity, $NH_3$, $NO_3$-N, $PO_4$-P | Utility |
| | | DOC, UV, conductivity, T, pH, ELISA | CSM |
| | | pH, T, GC-MS; bioassays | WSLH |

## Facility I

### *Treatment*

Facility I is designed to treat 1.5-mgd and currently operates at 1-mgd. It receives approximately 95% domestic sewage and 5% industrial discharge. The service area has no combined sewers; therefore no stormwater runoff is treated. Drinking water supply in the service area of the facility is entirely based on local groundwater.

Treatment at Facility I consists of grit removal, a membrane bioreactor, and disinfection with ultraviolet light. The flow of treated effluent is locally used for landscape irrigation (10%) and habitat maintenance (90%). The process schematic of Facility I is illustrated in Figure A-9. The HRT during secondary treatment is approximately 5 hours.

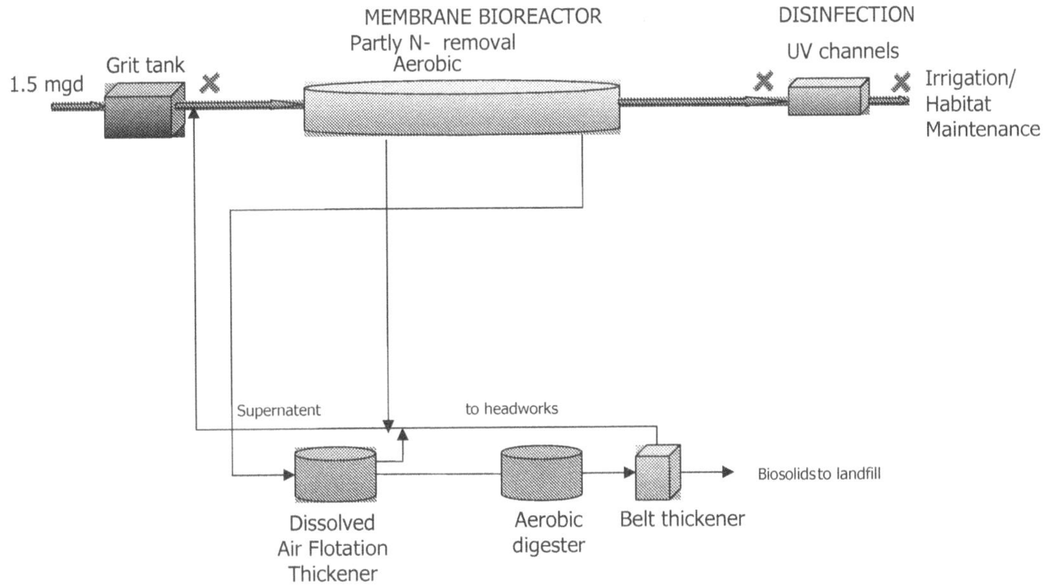


**Figure A-9. Process Schematic of Facility I (✖ Represents Water Quality Sampling Location).**

The facility employs on-site solids treatment sludge. Primary and excess biosolids are thickened prior to aerobic digestion. The digested biosolids are dewatered using a sludge press and hauled to a landfill. The supernatant of the digester is returned to the membrane bioreactor.

### *Water Quality*

Key operational parameters and water quality of secondary treated wastewater at Facility I are summarized in Table A-17 presenting monthly average values, minima and maxima for the time period 2001-2004.

Table A-17. Key Operational Parameters and Water Quality of Secondary Treated Effluent at Facility I.

| Parameter | Unit | Concentration |
|---|---|---|
| BOD loading | mg/L | 231<br>Min = N/A<br>Max = 261 |
| Mixed liquor total suspended solids | mg/L | 14,000 |
| HRT primary | Hrs | N/A |
| HRT secondary | Hrs | 5 |
| pH | - | 7.3<br>Min = N/A<br>Max = N/A |
| Conductivity | μS/cm | N/A |
| BOD | mg/L | 2.5<br>Min = N/A<br>Max = N/A |
| COD | mg/L | N/A<br>Min = N/A<br>Max = N/A |
| DOC | mg/L | 5.5<br>Min = N/A<br>Max = N/A |
| UV-254 absorbance | 1/m | 12<br>Min = N/A<br>Max = N/A |
| $NH_3$-N | mg/L | <0.5<br>Min = N/A<br>Max = N/A |
| $NO_3^-$-N | mg/L | 2.5<br>Min = N/A<br>Max = N/A |
| $PO_4^{3-}$-P | mg/L | <0.05<br>Min = N/A<br>Max = N/A |

***Sampling Locations***

A total of three composite samples were collected at Facility I examining the removal efficiencies of endocrine disrupting activity membrane bioreactor treatment followed by UV disinfection (Table A-18). A MBR influent composite sample were collected at the weir of the MBR, a secondary treated effluent sample was sampled at the effluent weir of the MBR, and a final effluent sample was collected after UV disinfection prior to discharge. The 4-hr composite samples at each unit operation considered the hydraulic retention time of each unit.

Table A-18. Sampling Locations and Target Parameters for Facility I.

| Sample Location | Type of Sample and Timing | Parameter | Responsible |
|---|---|---|---|
| MBR Influent | 4-hr composite<br>Time: 0 | DOC, UV, pH, T, conductivity, $NH_3$-N, $NO_3$-N, $PO_4$-P | CSM |
| | | UV, conductivity, T, pH, ELISA | CSM |
| | | pH, T<br>GC-MS; bioassays | WSLH |
| MBR Effluent<br>UV Influent | 4-hr composite<br>Time: 0 + 5 hrs | DOC, UV, pH, T, MLTSS, conductivity, $NH_3$, $NO_3$-N, $PO_4$-P | CSM |
| | | UV, conductivity, T, pH, ELISA | CSM |
| | | pH, T<br>GC-MS; bioassays | WSLH |
| UV Effluent | 4-hr composite<br>Time: 0 + 5.25 hrs | DOC, UV, pH, T, conductivity, MLTSS, $NH_3$, $PO_4$-P | CSM |
| | | UV, conductivity, T, pH, ELISA | CSM |
| | | pH, T<br>GC-MS; bioassays | WSLH |

## Facility J:

### *Treatment*

Facility J is designed to treat 40-mgd and currently operates at design capacity with an advanced water treatment capacity of 12-mgd. It receives approximately 90% residential sewage and 10% industrial discharge generated in the city. The service area has no combined sewers, therefore no stormwater runoff is treated. Drinking water supply in the service area of the facility is a blend of surface and groundwater. The groundwater originates from the local aquifer and represents approximately 20% of the drinking water supply.

Treatment at Facility J consists of grit removal, primary clarification, activated sludge treatment with nitrification/denitrification through a sequence of anoxic and aerobic zones, secondary clarification, tertiary dual-media filtration (anthracite, sand), disinfection using chloramines, optional microfiltration, and reverse osmosis. The process schematic of Facility J is illustrated in Figure F-10.

The hydraulic retention time during primary treatment is approximately 3.4 hrs at 12-mgd. The average mixed liquor aeration time is 8 hours and the solid retention time is approximately 9 days. The average detention time in the dual-media filters is 30 minutes.

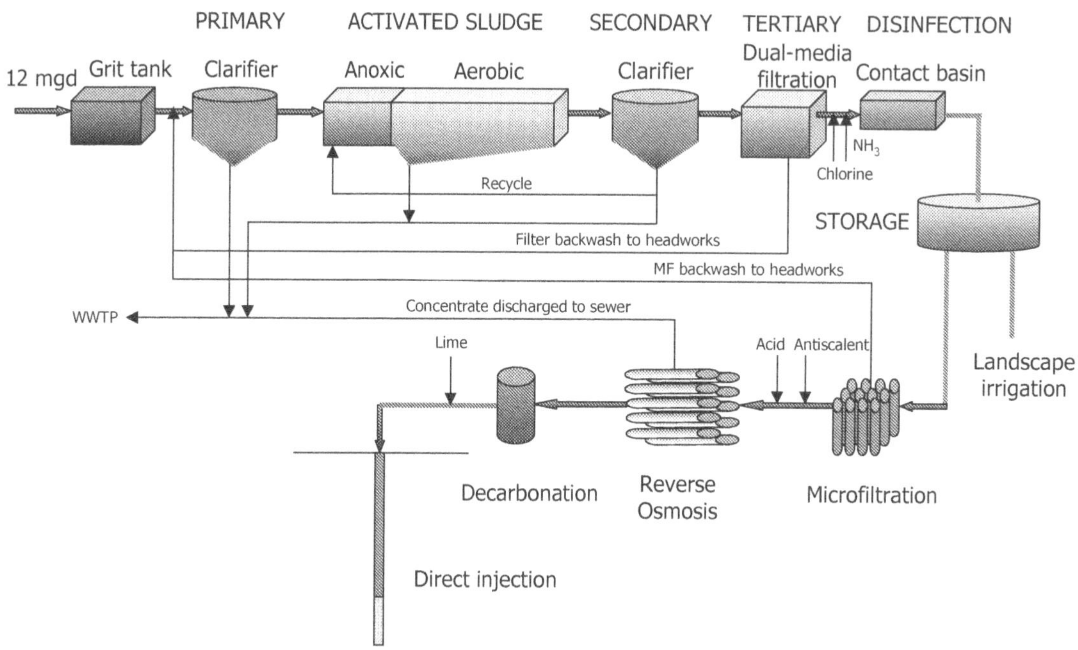


**Figure A-10. Process Schematic of Facility J.**

The facility does not employ on-site solids treatment. Excess sludge is discharged and transported via sewer to a centralized treatment facility.

### *Water Quality*

Key operational parameters and water quality of secondary treated wastewater at Facility J are summarized in Table A-19 presenting results of key parameter based on monthly average values for the time period 2000-2002.

**Table A-19. Key Operational Parameters and Water Quality of Secondary Treated Effluent at Facility J.**

| Parameter | Unit | Concentration |
|---|---|---|
| BOD loading | mg/L | 230<br>Min = 202<br>Max = 251 |
| Mixed liquor total suspended solids | mg/L | 1,600<br>Min = 1,972<br>Max = 1,480 |
| HRT primary | Hrs | 3.4 |
| HRT secondary | Hrs | 8 |
| SRT | Days | 9 |
| pH | - | 7.5<br>Min = N/A<br>Max = N/A |
| Conductivity | µS/cm | 1450<br>Min = N/A<br>Max = N/A |
| BOD | mg/L | 2<br>Min = N/A<br>Max = N/A |
| COD | mg/L | 25<br>Min = 22<br>Max = 30 |
| DOC | mg/L | 7<br>Min = N/A<br>Max = N/A |
| UV-254 absorbance | 1/m | N/A |
| $NH_3$-N | mg/L | 1.4<br>Min = 1.2<br>Max = 1.9 |
| $NO_3^-$-N | mg/L | 8.5<br>Min = 7.7<br>Max = 10.1 |
| $PO_4^{3-}$-P | mg/L | 2.4<br>Min = 2.4<br>Max = 2.6 |

### *Sampling Locations*

Sampling at Facility J occurred prior to and after an individual RO train.

**Facility K:**
**Treatment**

Facility KF is designed to treat 15-mgd and currently operates at 10-mgd. It receives approximately 90% residential sewage and 10% industrial discharge. The service area has no combined sewers, therefore no stormwater runoff is treated. Drinking water supply in the service area of the facility is a blend of surface and groundwater.

Treatment at Facility K consists of primary clarification, activated sludge treatment with nitrification/denitrification through a sequence of anoxic and aerobic zones, secondary clarification, alum addition for coagulation prior to filtration, disinfection using sodium hypochlorite (pre-filter chlorination is used), tertiary dual-media filtration (anthracite, sand), and chlorine contact basin. Ammonia is added prior to chlorination to achieve chloramination. Sodium bisulfite is used after chlorine contact to dechlorinate prior to discharge. The process schematic of Facility K is illustrated in Figure A-11.

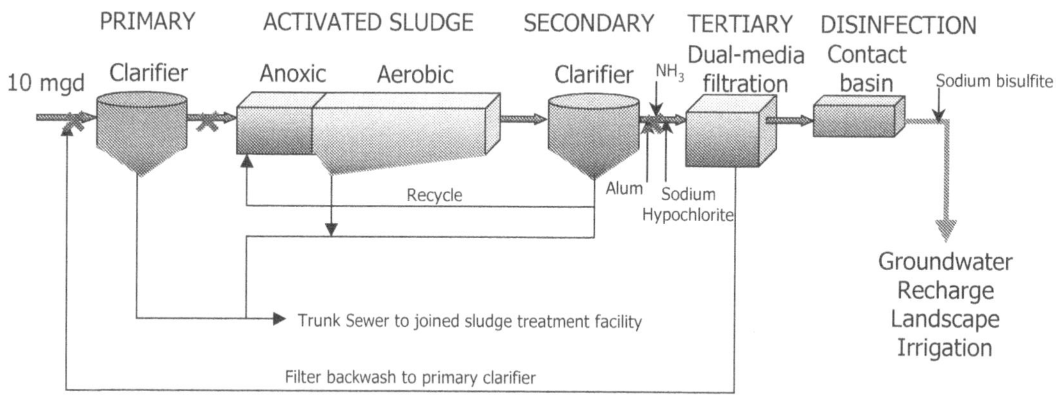


**Figure A-11. Process Schematic of Facility K (✗ Represents Water Quality Sampling Locations).**

The hydraulic retention time during primary treatment is approximately 3.2 hrs at 10-mgd. The average mixed liquor aeration time is 3.2 hours and the solid retention time is approximately 19 days. The average detention time in the dual-media filters is 30 minutes. The contact time in the chlorination contact basin is approximately 90 minutes. For fiscal year 2001-2002, 98.3% of the effluent produced at Facility K was actively reused for groundwater recharge and irrigation purposes. The great majority (99.2%) of the reclaimed water was used to recharge the groundwater aquifer. The remaining 0.8% of reclaimed water was used for ornamental plant irrigation at a local nursery.

The facility does not employ on-site solids treatment. Excess sludge is discharged and transported via sewer to a centralized treatment facility.

***Water Quality***

Key operational parameters and water quality of secondary treated wastewater at Facility K are summarized in Table A-20 presenting monthly average values, minima and maxima for the time period 1997-2002. The effluent quality meets California Title 22 standards for recycled water.

Table A-20. Key Operational Parameters and Water Quality of Secondary Treated Effluent at Facility K.

| Parameter | Unit | Concentration |
|---|---|---|
| BOD loading | lbs/day | 11,000<br>Min = 6,200<br>Max = 15,000 |
| Mixed liquor total suspended solids | mg/L | 3,400<br>Min = 1,760<br>Max = 3,500 |
| HRT primary | Hrs | 3.2 |
| HRT secondary | Hrs | 3.2 |
| SRT | Days | 19 |
| pH | - | 7.3<br>Min = 7.0<br>Max = 7.4 |
| Conductivity | μS/cm | 940<br>Min = 890<br>Max = 1,120 |
| BOD | mg/L | 2<br>Min = 1.1<br>Max = 4 |
| COD | mg/L | 23<br>Min = 18<br>Max = 31 |
| DOC | mg/L | 7.3<br>Min = 6.3<br>Max = 12.8 |
| UV-254 absorbance | 1/m | N/A |
| $NH_3$-N | mg/L | 1.2<br>Min = 0.1<br>Max = 3.8 |
| $NO_3^-$-N | mg/L | 6.2<br>Min = 4.0<br>Max = 7.8 |
| $PO_4^{3-}$-P | mg/L | 2.6<br>Min = 0.7<br>Max = 4.7 |

### *Sampling Locations*

For background comparison, a surface water sample was collected from the surface water source (Table A-21). Three composite samples were collected at Facility K to examine the removal efficiencies of endocrine disrupting activity during primary and activated sludge treatment. A raw sewage composite sample was collected prior to the primary clarifier. A primary effluent sample at the weir of the primary clarifier and a secondary treated effluent at the weir of the secondary clarifier prior to chlorination were collected. The 4-hr composite samples at each unit operation considered the hydraulic retention time of each unit.

Table A-21. Sampling Locations and Target Parameters for Facility K.

| Sample Location | Type of Sample and Timing | Parameter | Responsible |
|---|---|---|---|
| Surface Water | Grab sample | pH, T, dissolved oxygen | Utility, WSLH |
| | | DOC, UV, conductivity, T, pH, ELISA | CSM |
| | | pH, T<br>GC-MS; bioassays | WSLH |
| Raw sewage prior to primary clarifier | 4-hr composite<br>Time: 0 | pH, T, conductivity, $NH_3$-N, $PO_4$-P | Utility |
| | | DOC, UV, conductivity, T, pH, ELISA | CSM |
| | | pH, T<br>GC-MS; bioassays | WSLH |
| Primary effluent at weir of primary clarifier | 4-hr composite<br>Time: 0 + 3.5 hrs | pH, T, conductivity, $NH_3$-N, $NO_3$-N, $PO_4$-P | Utility |
| | | DOC, UV, conductivity, T, pH, ELISA | CSM |
| | | pH, T<br>GC-MS; bioassays | WSLH |
| Secondary effluent at weir of secondary clarifier | 4-hr composite<br>Time: 0 + 7 hrs | pH, T, MLTSS, conductivity, $NH_3$, $NO_3$-N, $PO_4$-P | Utility |
| | | DOC, UV, conductivity, T, pH, ELISA | CSM |
| | | pH, T<br>GC-MS; bioassays | WSLH |

**Facility L:**

***Treatment***

Facility L is designed to treat 175.9-mgd and currently operates at design capacity. The exact ratio of domestic to industrial sewage is not known. The service area has no combined sewers, therefore no stormwater runoff is treated. Drinking water supply in the service area of the facility is a blend of surface and groundwater. The groundwater originates from the local aquifer. The majority of the drinking water supply is surface water.

Treatment at Facility L consists of grit removal, primary clarification, activated sludge treatment with nitrification/denitrification through a sequence of anoxic and aerobic zones, secondary clarification, and disinfection using chlorine. The process schematic of Facility L is illustrated in Figure A-12. The hydraulic retention time during primary treatment is approximately 1.5 at 175.9-mgd. The average mixed liquor aeration time is 5 hours and the solid retention time varies between 6 and 11 days. The majority of the treated effluent is discharged to a river. Part of this effluent flow is diverted to a wetlands demonstration project.

The facility employs on-site solids treatment for primary and excess activated sludge. On-site biosolids treatment consists of thickening centrifuges for primary sludge and dissolved air flotation for biosolids prior to mesophilic anaerobic digestion with centrifuge dewatering. The centrates are returned to the headworks.

### *Water Quality*

Key operational parameters and water quality of secondary treated wastewater at Facility L are summarized in Table A-22 presenting monthly average values, minima and maxima for the time period 1997-2002.

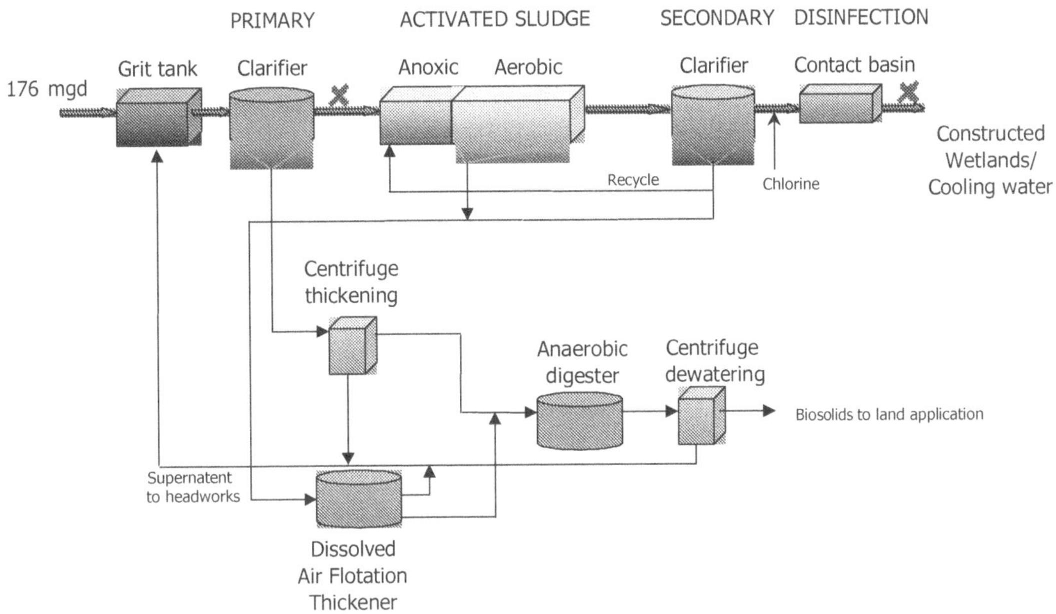


**Figure A-12. Process Schematic of Facility L ( ✗represents water quality sampling locations).**

Table A-22. Key Operational Parameters and Water Quality of Secondary Treated Effluent at Facility L.

| Parameter | Unit | Concentration |
|---|---|---|
| BOD loading | mg/L | 296<br>Min = 264<br>Max = 327 |
| Mixed liquor total suspended solids | mg/L | 3,290<br>Min = 3,038<br>Max = 3,415 |
| HRT primary | Hrs | 1.5 |
| HRT secondary | Hrs | 5 |
| SRT | Days | 6-11 |
| pH | - | 7.3<br>Min = 7.03<br>Max = 7.39 |
| Conductivity | μS/cm | N/A |
| CBOD | mg/L | 3<br>Min = 2<br>Max = 4 |
| COD | mg/L | N/A |
| DOC | mg/L | N/A |
| UV-254 absorbance | 1/m | N/A |
| $NH_3$-N | mg/L | 2.0<br>Min = 2.0<br>Max = 2.0 |
| $NO_3^-$-N | mg/L | 4.0<br>Min = 3.6<br>Max = 4.1 |
| $PO_4^{3-}$-P | mg/L | N/A |

***Sampling Locations***

Two composite samples were collected at Facility L to examine the removal efficiencies of endocrine disrupting activity during secondary treatment (nitrification/denitrification) followed by chlorine disinfection (Table A-23). A primary effluent sample at the weir of the primary clarifier and a secondary treated effluent after the chlorine contact basin were sampled. The final 4-hr composite sample considered the hydraulic retention time of all process operations.

Table A-23. Sampling Locations and Target Parameters for Facility L.

| Sample Location | Type of Sample and Timing | Parameter | Responsible |
|---|---|---|---|
| Primary effluent at weir of primary clarifier | 4-hr composite<br>Time: 0 hrs | pH, T, conductivity, $NH_3$-N, $NO_3$-N, $PO_4$-P | ASU |
| | | DOC, UV, conductivity, T, pH, ELISA | CSM |
| | | pH, T<br>GC-MS; bioassays | WSLH |
| Final effluent after disinfection | 4-hr composite<br>Time: 0 + 8 hrs | pH, T, conductivity, $NH_3$, $NO_3$-N, $PO_4$-P | ASU |
| | | DOC, UV, conductivity, T, pH, ELISA | CSM |
| | | pH, T<br>GC-MS; bioassays | WSLH |

# APPENDIX B

# SAMPLE COLLECTION PROTOCOL

## *Wisconsin State Laboratory of Hygiene*
## *Aquatic Life Toxicity Testing Laboratory*

---

### SAMPLE COLLECTION FOR BIOASSAY AND CHEMICAL ANALYSIS FOR ENDOCRINE DISRUPTION ACTIVITY

**WI State Lab of Hygiene**
**Biomonitoring Lab**
Contacts:
Jocelyn Hemming
(hemminjc@mail.slh.wisc.edu)
Miel Barman
(mabarman@mail.slh.wisc.edu)

Phone: (608) 224-6230
Fax: (608) 226-6267

Shipping Address:

Biomonitoring
2601 Agriculture Drive
Madison, WI 53718

**Colorado School of Mines**
Contacts:
Jörg Drewes
(303)273-3401
(jdrewes@Mines.EDU)
Sarah Ladenburger
(303)384-2445
(sladenburger@yahoo.com)

Shipping Address:

Jörg Drewes, Colorado School of Mines
Environmental Science & Engineering Division
Central Stores and Receiving
900 18th Street, Lower Level
Golden, CO 80401-1887

1. **Description and Application**

This method describes the sampling of effluents for subsequent bioassay and chemical analysis for endocrine disruption activity performed at WSLH and CSM. Because many plastics leach chemicals that can interfere with analyses, every effort is made to avoid contact of the sample with plastics. Occasionally use of plastics is unavoidable and a sampling blank is used to identify compounds that are present due to sampling manipulations.

2. **Equipment Preparation**

Equipment that will come in direct contact with the samples must be acid-washed using the following procedure in the order described.

A. Wash each item with hot water/soap solution and rinse in tap water.
B. Rinse with 10% hydrochloric acid (v:v), (10% nitric acid may be substituted).
C. Rinse with DI water (Type II water).
D. Rinse with methanol. Rinsing should be done under a hood to prevent the inhalation of fumes.
E. Rinse three times with distilled-quality water (Type I water). (This water should NOT be stored in a plastic container or squirt bottle)

3. **Equipment**

**A. Equipment needed to collect effluent.**

1) Composite sampler
2) Tubing and tubing connectors (we use Fisher cat # 14-168-10R for pump, Fisher cat # 14-169-7H used to connect pump tube to water sample) (new tubing should be used at each site—tubing is not cleaned, only rinsed with sample prior to use.)
3) Knife to cut tubing
4) Ice (20 lbs. per cooler)
5) Large enough glass container (probably > 6L) to collect sample (cleaned with above protocol)
6) Coolers
7) Lab Slips
8) Gloves
9) Sample sieve with hose clamp
10) Graduated cylinder (500 mL) for sampler calibration
11) Nylon tie
12) Thermometer
13) pH meter or paper
14) Sulfuric Acid (conc., HPLC grade)
15) Bottles to collect samples for facility and ASU

**B. Equipment needed to filter samples**

1) Large glass container (cleaned with above protocol— to collect filtered sample)

2) Glass fiber filters (e.g. Whatman GF—muffled at 450°C for 4 hours or rinsed with dichloromethane, followed by methanol then Type I water.)
3) Filtering apparatus (all components cleaned with above protocol) (e.g. Buchner funnel, vacuum flask)

**C. Equipment needed to ship samples**

1) 1 liter amber bottles (5 per sample)
2) Shipping kit (1 per 6 bottles)
3) Ice (10 lbs per kit) or ice packs
4) Packing material (e.g. shipping peanuts or bubble wrap)

## 4. Sample Integrity

It is imperative that the sample integrity is maintained. Care must be taken to keep samples free from contamination and cool prior to and during shipment to the laboratory. It is the intention of the analyses to test the sample as near to original composition as possible. Effluents should be subjected to minimal aeration during collection, filtration and storage to prevent oxidation and the loss of volatile compounds.

## 5. Collection

### Collection of effluent

A 4-hour composite sample is collected using a composite sampler. If one is not available, the composite can be made from grab samples taken at timed intervals and manually mixed.

1) If the sample is to be collected with an automatic sampler, follow the manual instructions for operating the sampler.
2) For a 4-hour composite sample, calculate the volume needed for samples every 10 minutes. (E.g., 250 mL x 6/hour x 4 hours = 6 L).
3) New tubing (not pre-cleaned), should be installed prior to sampling a new site.
4) Connect sample sieve to sample tubing and submerge in sample being collected.
5) Calibrate the sampler using the graduated cylinder.
6) Record temperature and pH of sample on lab slip. Water wasted during calibration is ideal for measuring these.
7) Rinse all sampling equipment (glass container, tubing, etc.) with sample water before beginning the sampling.
8) Label the large pre-cleaned, rinsed glass container with the sample number.
9) Center the empty glass container in the cooler. Pack approximately 20 lbs. of cubed ice around the glass container.
10) Insert the tubing into the glass container no more than two inches to avoid sample being removed during the sampler's purge cycle.
11) Record time of sample initiation on lab slip.
12) After the 4-hour collection, record ending time of sample.

13) Mix sample well then pour off water sample as requested by the facility for testing.
14) Acidify the remaining water sample to pH 2-3 using clean sulfuric acid (conc.) (pour off sample to test, do not put pH paper or probe in entire sample).
15) Pack cooler with additional ice as needed to keep cool during transport back to lab.

**BLANKS, (sample and travel)**

To determine contamination from the sample manipulations, a sample blank will be collected using Type I water from WSLH, shipped prior to sampling.

1) Empty the 4 L of Type I water provided in the 1 L amber bottles into a large pre-cleaned glass container. (Save the empty bottles and use to ship the samples to CSM and WSLH).
2) Use techniques mimicking collection of samples (steps 1-15 above). Eliminating collection of sample for facility and ASU.
3) A 1 L travel blank was also included, add acid to this bottle and return to WSLH.

**Lab Slip**

During each sample collection the Lab Slip should be filled out. Include any deviations from protocols or unusual events.

**Filtering samples**

1) All water samples (including the sampling blank) must be filtered through a pre-cleaned glass fiber filter (0.45 um).
2) Filter water into pre-cleaned glass container.

**Sample splitting and shipping**

1) After filtration, mix water sample to ensure homogeneity.
2) Label the 1 L amber bottles to include Facility ID (e.g. Facility C), sample location (e.g. 1°clarifier), date of collection, and indicate that samples are filtered and acidified. 3 1-L bottles are shipped to Colorado School of Mines and 2 1-L bottles are shipped to Wisconsin State Laboratory of Hygiene for each sample location.
3) For the sampling blank, 2 L are shipped to WSLH and 2 L are shipped to CSM. The travel blank is not processed after the acid addition and is returned to WSLH.
4) Pour appropriate samples into 1 liter glass amber bottles.
5) Bottles are placed into shipping kits. Pack kit with ice or ice packs and fill any remaining space with packing material. Tie bag shut, include lab slips and seal box shut with tape.
6) Ship via FedEx for overnight delivery. It appears as though writing FRAGILE LIVE FISH on the package helps prevent mishaps in handling that result in bottle breakage.

Address the samples to:

| Jörg Drewes<br>Environmental Science & Engineering Division<br>Colorado School of Mines<br>Central Stores and Receiving<br>900 18th Street, lower level<br>Golden, CO 80401-1887<br>Phone 303-273-3401 | Biomonitoring Lab<br>Wisconsin State Lab of Hygiene<br>2601 Agriculture Dr.<br>Madison, WI 53718 |
|---|---|

APPENDIX C

# ACID PRESERVATION OF SAMPLES

## Acid Preservation of Samples

The research team explored using acid to preserve the samples and the use of deutcrated cholesterol as an internal standard. At Facility B they collected enough sample that they could split the sample. Two thirds of the sample was preserved with sulfuric acid to pH 2-3. The remaining third was untreated. A portion of the preserved sample was spiked with deuterated cholesterol.

**Table C-1. Comparing Results for Preserved and Not Preserved Samples.**

| | Primary treated effluent | | | Secondary treated effluent | | |
|---|---|---|---|---|---|---|
| **Standards GC-NCI-MS** | **Acidified** | **Acidified + cholesterol** | **Not preserved** | **Acidified** | **Acidified + cholesterol** | **Not preserved** |
| 17α-ethynylestradiol | 2.2 | 4.6 | 9.1 | <0.2 | 0.3 | 0.3 |
| 17β-estradiol | 10.7 | 15.1 | 8.9 | <1 | <1 | <1 |
| Estriol | 206 | 180 | 307 | <1 | 1.3 | <1 |
| Estrone | 35 | 41 | 42 | 5.7 | 6.6 | 6.6 |
| Testosterone | 12.4 | 27.3 | 27 | <1 | <1 | <1 |
| Nonylphenol | 84,600 | 43,900 | 192,000 | 2,010 | 2,110 | 4,870 |
| 4-t-Octylphenol | 23,200 | 11,700 | 27,100 | 134 | 204 | 239 |
| 4-Octylphenol | < 50 | < 50 | < 50 | < 50 | < 50 | < 50 |
| Bisphenol A | 140,000 | 106,000 | 150,000 | 3,250 | 2,940 | 3,250 |
| **ELISA** | | | | | | |
| Estriol | 29.6 | 48.3 | 19.2 | 1.97 | 1.86 | <0.6 |
| 17β-estradiol | 12.7 | 9.73 | 37.4 | <0.6 | 2.6 | <0.6 |
| Testosterone | 36.2 | 47.1 | 21.9 | 1.07 | 1.60 | <0.7 |
| **Bioassay** | | | | | | |
| E-screen | 64.4 | 53.1 | 60.3 | 1.22 | 1.38 | 2.04 |
| ER-yeast | 23.2 | 17.1 | 33.9 | 5.47 | 5.35 | 6.76 |
| A-screen | 312 | 394 | 711 | 11.3 | 9.72 | 12.5 |
| AR-yeast | 211 | 255 | 216 | <218 | <218 | <218 |

Overall, it was unclear that sample preservation improved recoveries of these compounds (Table C-1). Variability was high, and in several cases unpreserved samples indicated higher concentrations. It is also the case that the samples were extracted shortly after collection < 24 hours at WSLH and < 48 hours at CSM, and were kept cold, therefore it is not clear whether preservation would be as important under these ideal conditions. However, the researchers did continue to acidify samples based on the recommendations of a PSC member. They were unable to use cholesterol as an internal standard because it did not appear to derivatize for GC-NCI-MS analysis. The deuterated cholestrol did prove unreactive in the E-screen bioassay. The data from this facility (B) and the original sampling of Facility A were not used in the report because of several inconsistencies in the analytical approach that were corrected in the following samples. For example, there was considerable salt contamination in the extracts from the salting procedure used, which interfered with GC-NCI-MS analyses. This was corrected by using a Type I water rinse prior to solvent elution. Additionally, the source for water in the derivatization procedure was contaminated with the phenols and a cleaner source was used for subsequent samples. Thus, the research team had less confidence in the results from the first two facilities.

# APPENDIX D

# FACILITY A SAMPLING

Facility A was sampled July 8 and 9, 2003 following the proposed sampling strategy developed from information summarized in the Utility Survey.

**Background**

Facility A is designed to treat 50-mgd and currently operates at 42-mgd. It receives approximately 63% domestic sewage from a metropolitan area and 37% industrial discharge including wastewater from a major food processing company. The service area has no combined sewers, therefore no stormwater runoff is treated. Drinking water supply in the service area of the facility is entirely based on groundwater.

**Treatment and Hydraulics**

Treatment at Facility A consists of grit removal, primary clarification, activated sludge treatment, and ultraviolet disinfection prior to discharge. The raw wastewater enters the plant through four forcemains. These forcemains discharge into a structure that has three large pipes exiting it. Each of those conduits goes to one of three grit chambers. The grit chambers discharge into one channel, which has two conduits exiting it. One conduit conveys water to the East Plant and the other takes water to the West Plant, each operates at 21-mgd (80,000 $m^3$/day). About half of the entire flow entering the plant comes from Pump Station #7. This station collects water from the north and east portion of the collection system. It is believed that all the water entering Grit Tank #3 is water from Pump Station #7. Some of the flow from Pump Station #7 also enters Grit Tank #2. About 60% of the total plant flow enters the West Plant. Most of the flow is from Pump Station #7, but some of the flow is from the other pump stations.

After careful consultation with the plant operators, the team decided to collect a representative raw water sample of the facility at the channel leading to Grit Tank #3. An additional sample was collected after Grit Tank #3. The difference between these two samples are that the "before" sample will probably contain water only from Pump Station #7 while the "after" sample will also contain some water from other pumping stations. More importantly, the "after" sample also contains the supernatant return flows from the onsite biosolids treatment which most likely contain high amounts of endocrine disrupting activities.

The east and west plants operate three and two, respectively, parallel activated sludge trains followed by secondary clarifiers. A junction structure after secondary treatment combines all flows from the East Plant and the West Plant after secondary clarifications prior to UV disinfection. The average hydraulic retention time (HRT) during primary treatment is approximately 2.22 hrs at 42-mgd (160,000 $m^3$/day). To assess the removal efficiency of mechanical treatment, an additional sample was collected at the effluent channel of the West primary clarifiers. The water flowing to this location should hydraulically correspond to the water sample collected "after the Grit Tank". The average mixed liquor aeration time during secondary treatment is 13.7 hours and the solid retention time (SRT) is approximately 10 days. The activated sludge treatment consists of three zones, - anaerobic, anoxic and aerobic to achieve biological phosphorus removal, nitrification and denitrification. To assess the biological treatment process, a secondary clarifier sample was collected at the weir. The water flowing to this location should be the same as that flowing to the "After the Grit Tank" and "Primary Clarifier Weir" samples. To assess the impact of UV disinfection on endocrine disrupting activity, samples were collected prior to and after UV exposure. The sample prior to was collected from the covered channel entering the UV building. The sample after UV treatment was collected at the end of the UV channel prior to leaving the UV building. A sample taken from the channel entering the UV building will contain water that was treated from the East Plant as well as the West Plant. Approximately 93% of the treated effluent is used to augment flow in Creek I and approximately 8% of the total effluent is discharged to Creek II.

The facility employs on-site biosolids treatment for primary sludge and wasted biosolids. Primary sludge is screened, degritted, and thickened by gravity prior to anaerobic digestion. The supernatant of the gravity sludge thickener (GST) is returned to the primary clarifiers of the East Plant. The digested sludge is further thickened by a gravity belt thickener (GBT). Wasted biosolids are thickened by dissolved air flotation (DAF) prior to anaerobic digestion. The supernatant of DAF and GBT are combined and returned to the grit tanks prior to primary clarification. A "Supernatant to Grit Tanks" sample was collected in the GBT building.

**Sampling at Facility A**

In order to assess the impact of different matrices on recovery and extraction efficiencies for target compounds, WSLH staff collected preliminary samples on June 10 and 11, 2003. Grab samples were collected from after the grit tank (sampling location #2) and after the secondary clarifier weir (sampling location #4). The main sampling of Facility A occurred during the second week of July 2003 (07/07 – 07/10). The sampling was conducted by staff from CSM and WSLH with support from utility staff of Facility A. Composite samples were collected at 8 locations throughout the plant: before the grit tank (#1), after grit tank (#2), at the primary clarifier weir (#3), at the secondary clarifier weir (#4), before the UV channel (#5), after the UV channel (#6), supernatant from the gravity sludge thickener (#7), and supernatant to the grit tank (#8). Sample locations are shown in Figure D-1.

Since the flow rate into the plant changes often minute to minute, instantaneous detention times are difficult to calculate. Therefore, the team decided to use the average flow rate from the previous hours to calculate the detention time for individual unit operations. Staff of Facility A assisted in calculating the average detention times of each unit operation sampled and provided them to the team prior to sampling. Samples were collected by following the same slug of water as it traveled though the plant. This was achieved by using the detention times of each of the unit processes and taking samples accordingly. 4-hour composite samples were collected by autosamplers for all samples except the supernatant grit tank. Samples were collected in 6-L glass containers. The sample collected from the supernatant grit tank represented a detention time of 14-20 days and according to the utility the composition does not have a lot of variability.

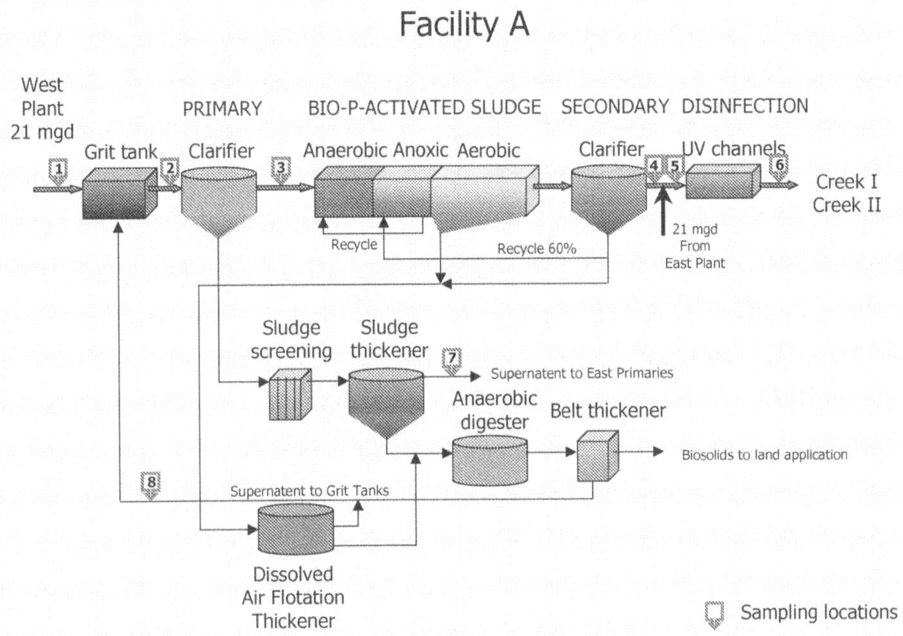


**Figure D-1. Sampling Locations at Facility A During Full-Scale Sampling.**

## Comparison of Results for Facility A Among Methods

### Comparison of Chemical Methods

Overall, chemical measurements and bioassays agreed reasonably well in assessing the general removal efficiencies of the treatment plant operations (Table D-1). At Facility A, EDCs were more markedly removed by the secondary than by primary treatment. Removal efficiencies by secondary treatment were generally greater than 90% for all compounds tested with all the methods. One exception was 17α-ethynylestradiol. The lower detection limits of the yeast assays precluded a good estimation of the removal of the activity.

Although the methods showed similar removal efficiencies, the concentrations obtained from the different methods were not always in agreement. For the two chemical methods (GC-NCI-MS and ELISA), measurements of testosterone were the most consistent (Figure D-2). With the exception of the unexpectedly low value by the GC-MS in the before grit tank sample from Facility A, overall agreement is generally good. Agreement between the 17β-estradiol results was not as consistent. Although the trends are similar, the actual values were 2.5-18 times higher for the ELISA results than for the GC-MS. The estriol results also indicate marked differences between the GC-MS and ELISA, with GC-MS producing the higher value in most cases.

Table D-1. Removal of EDCs Over Specific Processes as Assessed by GC-MS-NCI, ELISA and Bioassays.

| Method | Removal by Primary Treatment (%) | Removal by Secondary Treatment (%) | Total Removal (%) |
|---|---|---|---|
| **GC/MS/NCI** | | | |
| 17ß-estradiol | 57 | >86 | >94 |
| Estriol | 34 | >99 | >99 |
| Estrone | -14 | 99 | 98 |
| 17α-ethynylestradiol | 47 | 84 | >92 |
| Testosterone | 64 | >99 | >99 |
| Bisphenol A | 35 | >73 | >83 |
| **HPLC-ELISA** | | | |
| 17ß-estradiol | -141 | 94 | 85 |
| Estriol | -19 | >99 | >99 |
| Testosterone | 18 | 95 | 96 |
| **Bioassays** | | | |
| E-Screen | 33 | 99 | 99 |
| Yeast ER | -41 | >71 | >58 |
| Yeast AR | ND | ND | ND |

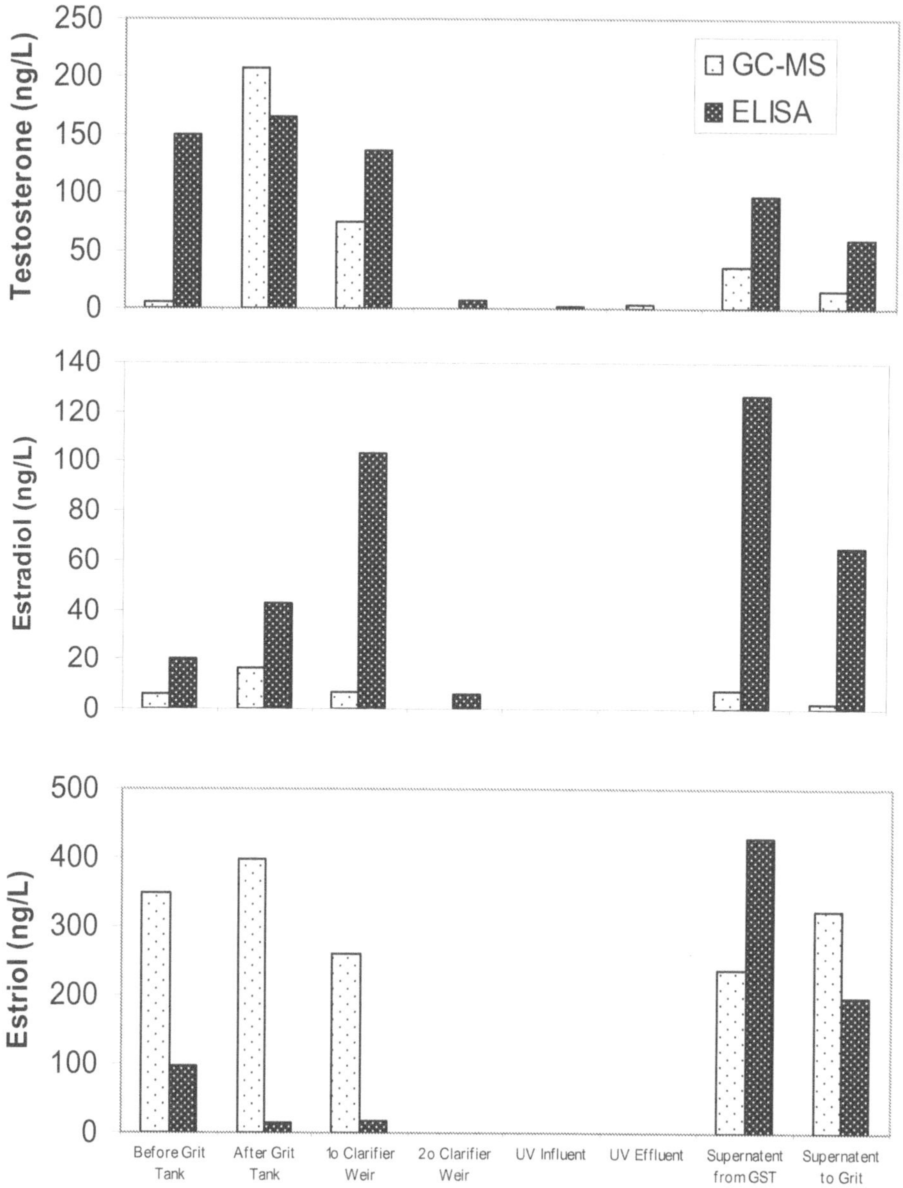


**Figure D-2. Comparison Between ELISA and GC-NCI-MS Results at Facility A.**

## Comparison of Estrogenic Activities

When the total activity of the chemicals is estimated by converting their concentration to 17β-estradiol equivalents, the chemical analyses can be compared to the bioassays. For the ELISA samples, the equivalents are based only on 17β-estradiol and estriol. The results from Facility A (Figure D-3) indicate that the chemical analyses would predict a much higher level of activity than was seen in the bioassays. This is especially true in the samples with complex matrices. Estriol dominates the activity for the "dirty" matrix samples representing untreated wastewater, accounting for about 100 ng/L of the estrogenic activity. The GC-MS analyses and E-screen results are in better agreement in the matrix representing treated wastewater samples (Figure D-4). The yeast assay and the E-Screen were quite similar in the magnitude of their responses, although the low detection limit of the yeast assays (approximately 6 ng/L) resulted in no detects for the samples collected after secondary treatment.

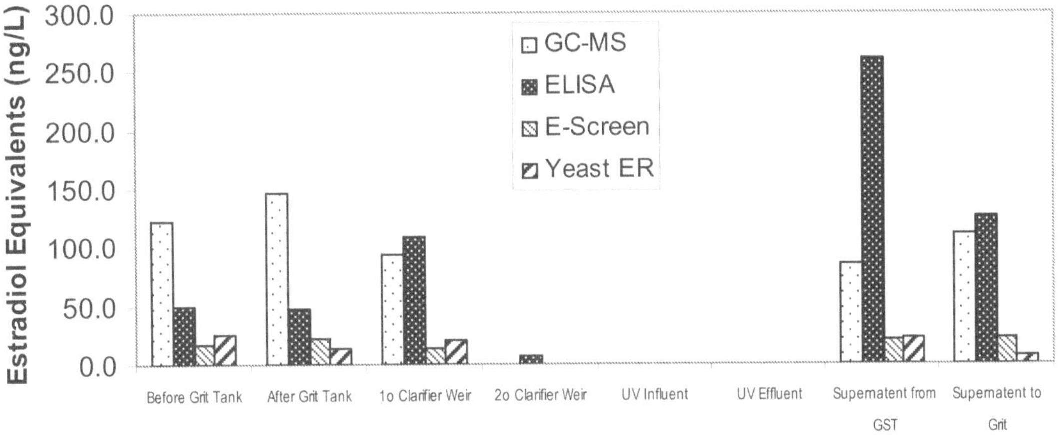


**Figure D-3. Results Expressed as Estradiol Equivalents at Facility A.**

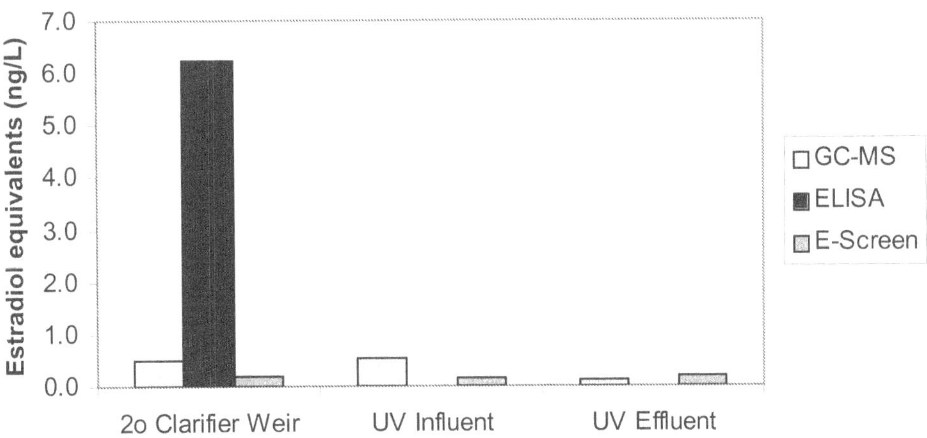


**Figure D-4. Results for Treated Samples Expressed as Estradiol Equivalents at Facility A.**

APPENDIX E

# EDC Changes during Treatment with Soil-Aquifer Treatment

Two full-scale facilities employing soil-aquifer treatment (SAT) were investigated during this study. Facility H is spreading tertiary treated effluent though recharge basins. Groundwater was sampled underneath the basin (1.5 m below surface) and after travel time of approximately 12 months. Only two EDCs, nonylphenol and estrone, were detected in the tertiary treated effluent. The effluent showed overall a low biological activity (Table E-1). Within a short distance of 1.5 m representing a travel time of approximately 5 days, estrogenic activity was further removed and no steroid hormone was detected. No EDC or activity was quantified in the monitoring well representing 12-months travel time.

**Table E-1. EDC Concentration and Activity Prior to and After SAT.**

| Compound | Tertiary treated effluent prior to spreading (ng/L) | Monitoring well 1.5 m (ng/L) | Monitoring well 12 months (ng/L) |
|---|---|---|---|
| Nonylphenol | 233 | 496 | n.d. |
| Estrone | 5 | n.d. | n.d. |
| E-screen | 0.67 | 0.11 | n.d. |
| ER-yeast | 1.44 | n.d. | n.d. |
| A-screen | 1.94 | n.d. | n.d. |
| AR-yeast | n.d. | n.d. | n.d. |

These findings were confirmed by controlled soil-column laboratory experiments where no EDCs were detected after simulated travel time of 30 days (Mansell et al., 2004). Findings were further supported by no detection observed at the second full-scale SAT site. These findings support that SAT is a very viable barrier for EDC removal.

# APPENDIX F

# ORGANIC MATTER INTERFERENCE IN THE YEAST ASSAYS

During sample analyses with the yeast assay, the research team sometimes found it difficult to quantify the activity because the dilutions of the samples (after being corrected for the dilution) yielded markedly different results. In the problematic samples, the different dilutions (before being corrected) resulted in similar β-galactosidase activities, thus the more dilute the sample, the higher the actual estimate of the activity. One possibility was that organic matter in the sample extract was interfering with the assay. To explore this possibility, a range of concentrations of EfOM extract (as generated in the spike recovery study) was added to the yeast assay with a constant hormone standard concentration. Additionally, a natural organic matter (NOM) standard (Suwannee River fulvic acid XAD-8 isolate IHSS, St. Paul, MN) was added at a range of concentrations to a separate series of assays with constant hormone standards.

As expected, the effluent organic matter contains estrogenic compounds, as indicated by the increase in activity with the increase in dose (Figure F-1). However, at a concentration corresponding to a 100-fold dilution of the sample in the assay, the EEq estimate was reduced to 67% of the control, even though the EfOM itself had measurable estrogenic activity. The natural organic matter samples, which should not contain hormones, did not indicate any estrogenic activity at the concentrations tested (Figure F-2). This organic matter also inhibited the activity of the ER-yeast assay, but at much higher concentrations (corresponding to a 10x-fold dilution of the sample in the assay).

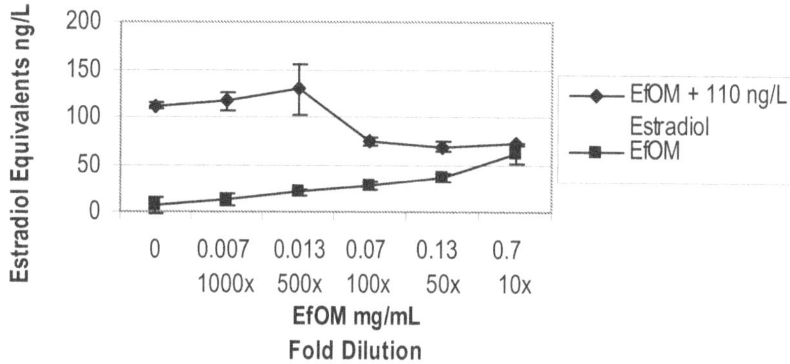


**Figure F-1. Influence of Effluent Organic Matter on ER-yeast Assay Spiked with 110 ng/L Estradiol.** The X-axis is the concentration of effluent organic matter as calculated from a 10 mg/L DOC EfOM solution concentrated with solid phase extraction and the corresponding x-fold dilution of the sample in the ER-yeast assay. Values represent means ± s.d. for two independent experiments.

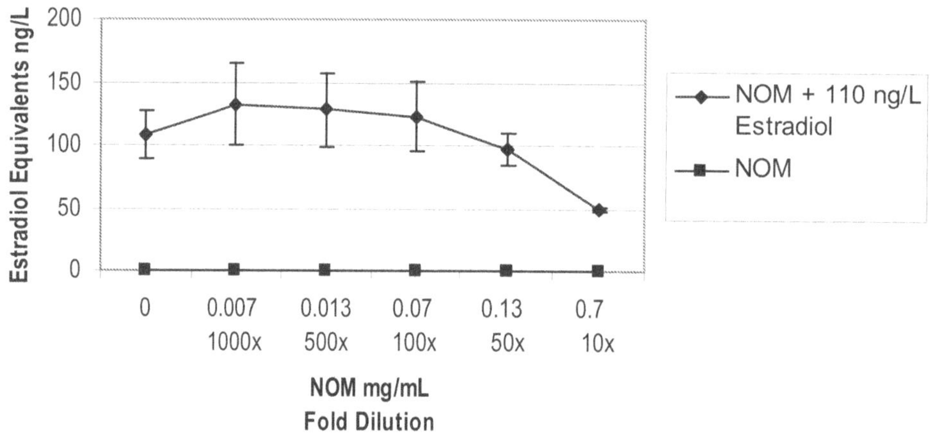


**Figure F-2. Influence of Natural Organic Matter on ER-Yeast Assay Spiked with 110 ng/L Estradiol.** The X-axis is the concentration of natural organic matter as calculated from a 10 mg/L DOC NOM solution concentrated with solid phase extraction and the corresponding x-fold dilution of the sample in the ER-yeast assay. Values represent means ± s.d. for two independent experiments.

In the AR-yeast assay, the EfOM did not have measurable androgen activity (Figure F-3). This is not suprising, as the sensitivity of this assay is more than an order of magnitude less than the ER-assay. The EfOM drastically reduced the activity similar to the ER-assay in that a 100-fold diltion of the sample caused a 54% reduction relative to controls. In contrast to the ER-assay, increasing the concentration of EfOM further diminished activity to levels lower than detection limits.

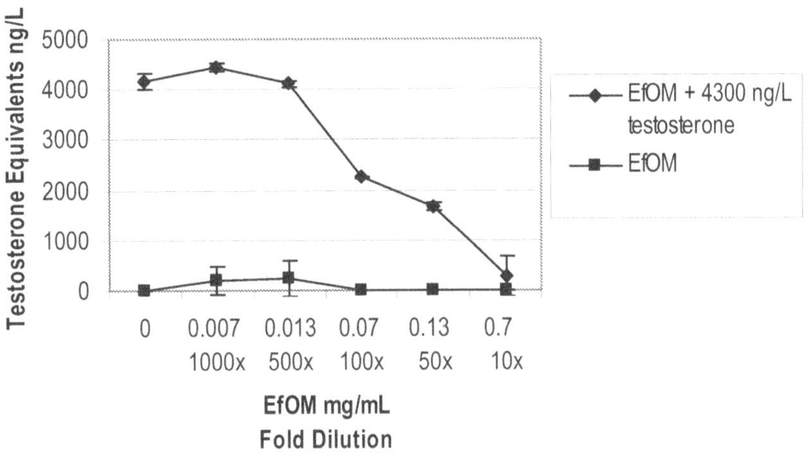


**Figure F-3. Influence of Effluent Organic Matter on AR-Yeast Assay Spiked with 4300 ng/L Testosterone.** The X-axis is the concentration of effluent organic matter as calculated from a 10 mg/L DOC EfOM solution concentrated with solid phase extraction and the corresponding x-fold dilution of the sample in the AR-yeast assay. Values represent means ± s.d. for two independent experiments.

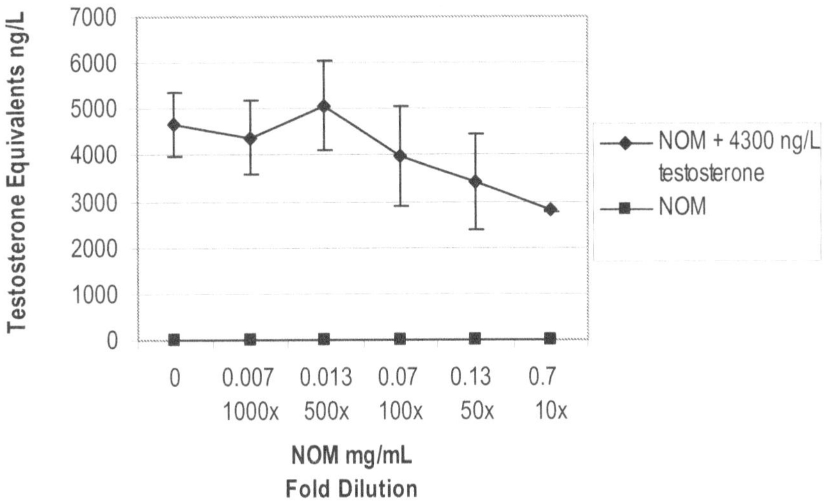


**Figure F-4. Influence of Natural Organic Matter on AR-Yeast Assay Spiked with 4300 ng/L Testosterone.** The X-axis is the concentration of natural organic matter and the corresponding x-fold dilution of the sample in the AR-yeast assay. Values represent means ± s.d. for two independent experiments.

The NOM also tended to decrease the activity of testosterone in the AR-yeast assay (Figure F-4), although similar to the NOM effects in the ER-yeast assay, the reduction was not as pronounced and occurred at the highest concentrations. Interestingly, the β-galactosidase activity in the standard control wells was sometimes negative in the prescence of high concentrations of natural organic matter, especially in the AR-yeast assay. The cause of this is unclear.

Yamamoto and Liljestrand (2004) recently reported the sorption coefficients of 17β-estradiol, 17α-ethinylestradiol, estriol, nonylphenol and octylphenol to a range of NOM using the fluorescence quenching technique. The sorption coefficients (Log $K_{oc}$ s) for binding to Suwannee River Fulvic acid were all between 4.55 and 4.70 and between 4.80 and 4.96 for Suwannee River Humic acid. Thus, a substantial proportion of these compounds can be expected to be bound to organic matter and therefore less bioavailable.

Tanghe et al. (1999) and Holbrook et al. (2005) also determined that the sensitivity of the yeast assay was decreased with samples containing humic acid and colloidal organic carbon, respectively, due to the binding of the estrogenic chemicals with the organic matter. Tashiro et al. (2004) and Iwasaki et al. (2004) suggest that yeast assay results can be improved by cleanup methods with HPLC. They found fractionating environmental samples prior to use can eliminate prolems in which the cytotoxicity of the extracts preclude the ability to indicate activity as well as removing compounds (such as organic matter) that decrease the bioavailability of the active components. It would be logical to expect that bioassays that are most sensitive (such as the E-screen) would be impacted by a lesser degree by organic matter interference because the dilutions of samples used are greater, but the least sensitive (such as the AR-yeast) would be most strongly affected by the interference.

# APPENDIX G

# Fish Vitellogenin Assay in MBR Effluents

## In Vivo Vitellogenin Assay

The County Sanitation Districts of Los Angeles County (Districts) conducted an in vivo fish bioassay to assess the endocrine disrupting activity in three different wastewater effluents. This project was conducted in conjunction with a series of *in vitro* assays performed by the Colorado School of Mines and Wisconsin State Laboratory of Hygiene. The Districts role in this project was to provide sampling support to perform a semi-quantitative vitellogenin Enzyme-Linked Immuno-Sorbent Assay (ELISA) on three types of effluent samples. A non-disinfected final effluent sample was collected from a Zenon and a Kubota Membrane Bioreactor (MBR) system. The Zenon MBR system treats an average daily flow of 30 gallons per minute (gpm) and has a nominal membrane filtration pore size of 0.04 micrometers (μm). The Kubota MBR system treats an average daily flow of 8.8 gpm and has a nominal membrane filtration pore size of 0.4 μm. In addition to the two MBR samples, a final effluent sample was collected from a Districts' full-scale tertiary treatment plant that provides for biological nitrogen removal, granular media filtration, chlorination and dechlorination.

The assay conducted was a 21-day fish bioassay, using juveniles (less than 30 days old) fathead minnow (*Pimephales promelas)* larvae from a commercial supplier, coupled with ELISA This method had previously been developed to examine source, wastewater and groundwater samples for the synthesis of vitellogenin (Vtg), an egg yolk precursor typically found in mature female fish, but can be induced in male fish when exposed to estrogen or estrogenic mimicking compounds. For each effluent sample, three fathead minnows were placed in six replicate beakers. The exposure vessels consisted of 400 mL glass beakers containing 300 mL of total solution. The fish were fed with daily renewal of test solutions (static-renewal). Upon test termination, the fish were homogenized and centrifuged. The plasma-containing supernatant was collected and frozen until ready for analysis by ELISA.

A carp vitellogenin (Vtg) ELISA kit was purchased from Biosense™ and the plasma-containing supernatant from the fish were analyzed for vitellogenin. The plasma vitellogenin level was converted into Vtg fold-induction relative to the mean control response. The Vtg fold-induction data was then transformed into 17α-ethinylestradiol equivalents (EE2Eq) obtained from historical 17α-ethinylestradiol equivalents reference tests modeled as: Vtg fold-induction versus the 17α-ethinylestradiol equivalents concentration. This EE2Eq conversion expresses the results in terms of the amount of 17α-ethinylestradiol equivalents required to produce similar levels of Vtg (estrogenic activity) in the fish exposed to the effluent samples. This conversion accounts for the variability observed in the results from background Vtg levels in the fish and/or the ELISA assay. Samples were determined to have significant estrogenic activity if the EE2Eq was equal to or greater than 14 pM (picomolar) and were statistically significant when compared to the associated control response. Samples where the EE2Eq were less than 14 pM and not significantly different from the control response were considered to have no estrogenic activity. Those samples that were suspect or potentially estrogenic had EE2 equivalents either greater than 14 pM and were not statistically significant or were less than 14 pM but were statistically significant from the control response.

The Zenon sample averaged 3.9 pM EE2Eq, but the average was not statistically significant from the average control response. There was no detectable level of estrogenic activity in the Kubota sample and the tertiary effluent sample averaged 0.56 pM EE2Eq, but was not statistically significant. Based on the semi-quantitative analysis of these results, none of the effluent samples exhibited estrogenic activity. A summary of the results is presented in Figure G-1.

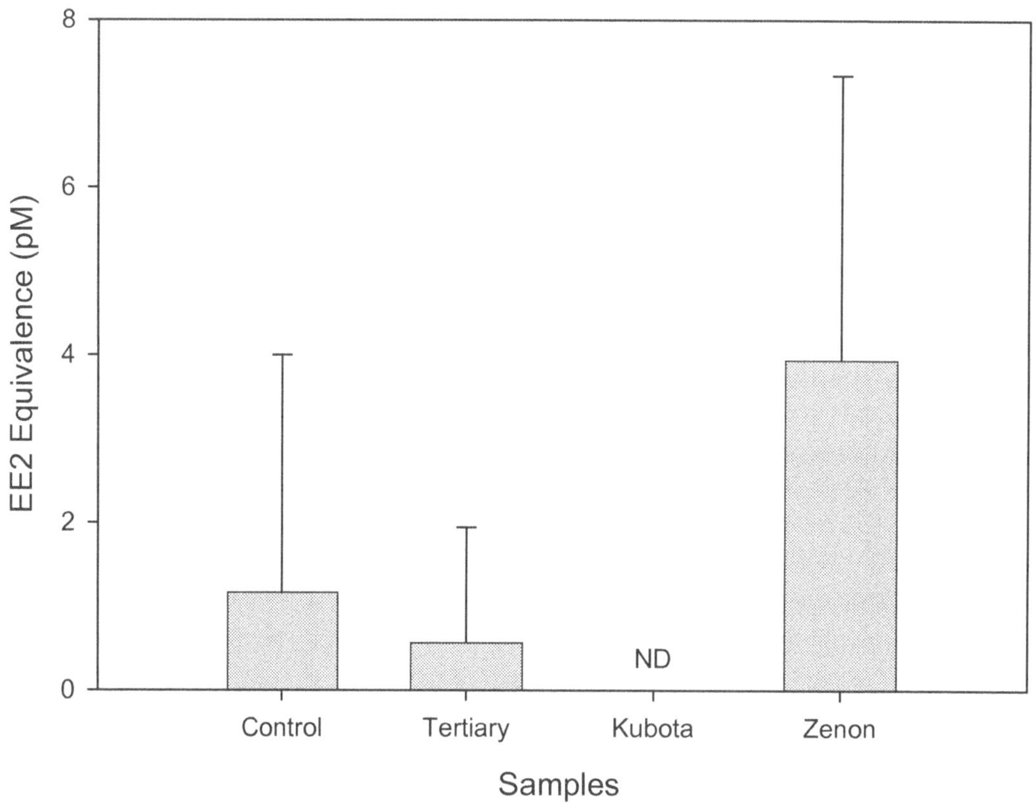


**Figure G-1. Induction of Vitellogenin in Fathead Minnow Juvenile Fish Exposed to Effluents After Tertiary Treatment or One of Two Membrane Bioreactors.** Values represent mean responses ± s.d. n = 6.

APPENDIX H

# EDC Changes during Treatment with Powdered Activated Carbon

Three powdered activated carbon adsorption isotherm experiments were conducted to determine the removal of EDCs by activated carbon (F300, Calgon Carbon). The carbon doses for the experiments ranged from 10 to 3,000 mg/L. The first experiment was conducted in Type I water (DI water) spiked with hormones and phenolic compounds. The second and third experiments were conducted in secondary effluent from Facility A spiked with hormones and phenolic compounds. The nominal spikes for the experiments are summarized in Table H-1.

**Table H-1. Nominal Spikes for Powdered Activated Carbon Experiments.**

| Compound | Experiment 1<br>Type I Water<br>Nominal Spike (ng/L) | Experiment 2<br>Secondary Effl.<br>Nominal Spike (ng/L) | Experiment 3<br>Secondary Effl.<br>Nominal Spike (ng/L) |
|---|---|---|---|
| Nonylphenol | 28,000 | 8,000 | 12,000 |
| 4-t-octylphenol | 7,000 | 2,000 | 3,000 |
| Octylphenol | 700 | 200 | 300 |
| Bisphenol A | 1,750 | 500 | 800 |
| 17β-Estradiol | 100 | 250 | 50 |
| Estrone | 100 | 250 | 50 |
| 17α-Ethinylestradiol | 40 | 250 | 20 |

After shaking the samples for 24 hours on an orbital shaker, the samples were analyzed for the target compounds. For experiment 1, none of the compounds were detected in the samples after shaking on an orbital shaker for 24 hours. In the presence of organic matter using a secondary treated effluent matrix (experiment 2), trace concentrations of nonylphenol and 4-t-octylphenol were observed for the lowest carbon dose. This may be due to the competition of the target compounds with the organic matter found in secondary effluent. When higher concentrations of phenolic compounds were used during spike experiments (experiment 3), nonylphenol was detected in samples with carbon doses ranging from 0 to 750 ng/L whereas no other EDC was detected. Results for experiment 3 nonylphenol concentrations are presented in Figure H-1.

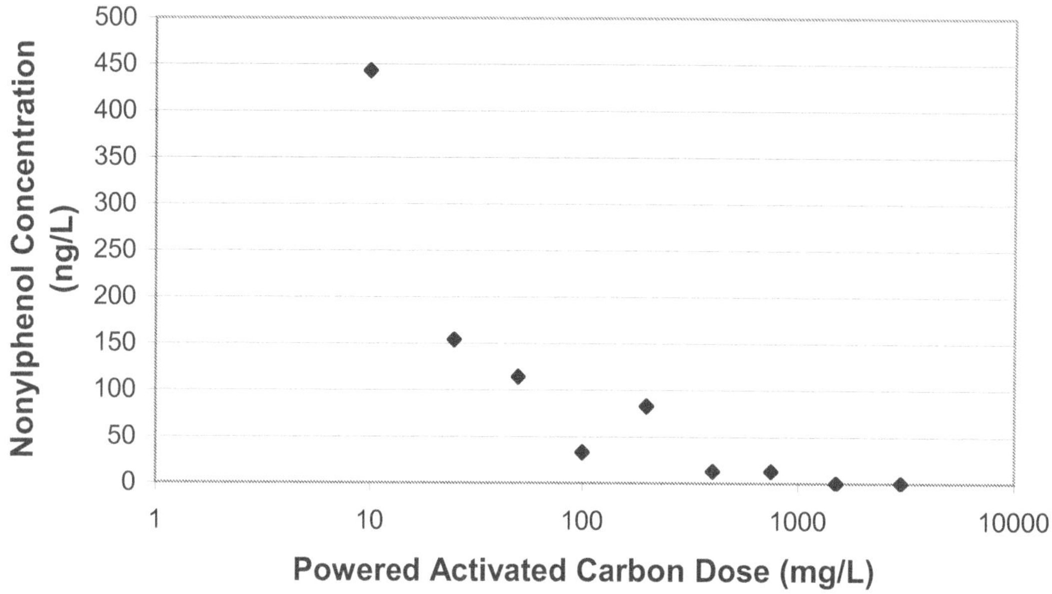


**Figure H-1. Nonylphenol Concentrations Versus Powered Actived Carbon Dose**

These results of the three experiments suggest that already a small amount of activated carbon added to secondary effluent may serve as an efficient and cost effective way to remove EDCs below detection limits. The purpose of this study was to assess the general adsorbability using rather long contact times (24 hrs), however, activated carbon is likely to be reasonable efficient at shorter and more realistic contact times. Adding powdered activated carbon to facilities that currently employ tertiary filtration would require minimal modifications of the treatment train, but possibly increase the amount of backwash needed.

# REFERENCES


American Water Works Association, American Public Health Association, Water Environment Federation (1998) *Standard Methods for the Examination of Water and Wastewater,* 20th edition. American Public Health Association, Washington, D.C.

Andersen, H.R., Andersson, A., Arnold, S.F. et al., (1999) Comparison of short-term estrogenicity tests for identification of hormone-disrupting chemicals. *Environ.Health Perspect.* 107(suppl 1), 89-108.

Andersen, H., Siegrist, H., Halling-Sørensen, B., and Ternes, T.A. (2003) Fate of estrogens in a municipal sewage treatment plant. *Environ. Sci. Tech.* 37(18), 4021-4026.

Aravindakshan, J., Paquet, V., Gregory, M., Dufresne, J., Fournier, M., Marcogliese, D.J., and Cyr, D.G. (2004) Consequences of xenoestrogen exposure on male reproductive function in spottail shinters (*Notropis hudsonius*). *Toxicol. Sci.* 78, 156-165.

Baronti, C., Curini, R., D'Ascenzo, G., Di Corcia, A., Gentili, A., and Samperi, R. (2000) Monitoring natural and synthetic estrogens at activated sludge sewage treatment plants and in a receiving river water. *Environ. Sci. Technol.* 34 (24), 5059-5066.

Belfroid, A., Van der Horst, A., Vethaak, A. D., Schäfer, A.J., Rijs, G.B.J., Wegener, J., and Cofino, W.P. (1999) Analysis and occurrence of estrogenic hormones and their glucuronides in surface water and waste water in The Netherlands. *Sci. Total Env.* 225, 101-108.

Belfroid, A., Van Velzen, M., Van der Horst, B., and Vethaak, D. (2002) Occurrence of bisphenol A in surface water and uptake in fish: evaluation of field measurements. *Chemosphere* 49, 97-103.

Bellona, C., Drewes, J.E., Xu, P., and Amy, G. (2004) Factors affecting the rejection of organic solutes during NF-RO treatment – A literature review. *Wat. Res.* 38, 2795-2809.

Bergeron, J.M., Willingham, E., Osborn, C.T. III, Rhen, T., and Crews, D. (1999) Developmental synergism of steroidal estrogens in sex determination. *Environ. Health Perspect.* 107(2), 93-97.

Birkett, J. and Lester, J. (2003) *Endocrine Disrupters in Wastewater and Sludge Treatment Processes.* IWA Publishing, Lewis Publishers CRC Press LLC.

Brian, J.V., Harris, C.A., Scholze, M., Backhaus, T., Booy, P., Lamoree, M., Pojana, G., Jonkers, N., Runnalls, T., Bonfà, A., Marcomini, A., and Sumpter, J.P. (2005) Accurate prediction of the response of freshwater fish to a mixture of estrogenic chemicals. *Environ. Health Perspect.* 113(6), 721-728.

Carballa, M., Omil, F., Lema, J. M., Llompart, M., García-Jares, C., Rodríguez, I., Gómez, M., and Ternes, T. (2004) Behavior of pharmaceuticals, cosmetics and hormones in a sewage treatment plant. *Water Research* 38, 2918-2926.

Clara, M., Kreuzinger, N., Streen, B., Gans, O., and Kroiss, H. (2005) The solids retention time–a suitable design parameter to evaluate the capacity of wastewater treatment plants to remove micropollutants. *Water Research* 39, 97-106.

Clara, M., Strenn, B. Saracevic, E., and Kreuzinger, N. (2004) Adsorption of bisphenol-A, 17β-estradiole, and 17α-ethinylestradiole to sewage sludge. *Chemosphere*. 56. 843-851.

Coleman, H.M., Abdullah, M.I., Egging, B.R., and Palmer, F.L. (2005) Photocatalytic degradation of 17β-oestradiol, oestriol, and 17α-ethynyloestradiol in water monitored using fluorescence spectroscopy. *Appl. Catalysis B: Environ*. 55, 23-30.

D'Ascenzo, G., Di Corcia, A., Gentili, A., Mancini, R., Mastropasqua, R., Nazzari, M., and Samperi, R. (2003) Fate of natural estrogen conjugates in municipal sewage transport and treatment facilities, *Sci. Total Env*. 302, 19-209.

Degitz, S.J., Holcombe, G.W., Flynn, K.M., Kosian, P.A., Korte, J.J., and Tietge, J.E. (2005) Progress towards development of an amphibian-based thyroid screening assay using Xenopus laevis. Organismal and thyroidal responses to the model compounds 6-propylthiouracil, methimazole, and thyroxine. *Toxicol. Sci*. 87(2), 353-364.

Desbrow, D., Routledge, E.J., Brighty, G.C., Sumpter, J.P., and Waldock, M. (1998) Identification of estrogenic chemicals in STW effluent. 1. Chemical fractionation and in vitro biological screening. *Environ. Sci. Technol*. 32(11), 1549-1558.

DeVito, M., Biegel, L., Brouwer, A., Brown, S., Brucker-Davis, F., Cheek, A.O., Christensen, R., Colborn, T., Cooke, P., Crissman, J., Crofton, K., Doerge, D., Gray, E., Hauser, P., Hurley, P., Kohn, M., Lazar, J., McMaster, S., McClain, M., McConnell, E., Meier, C., Miller, R., Tietge, J., and Tyl, R. (1999) Screening methods for thyroid hormone disruptors. *Environ. Health Perspect*. 107(5), 407-415.

Ding, W-H., and Wu, C-Y. (2000) Determination of estrogenic nonylphenol and bisphenol A in river water by solid-phase extraction and gas chromatography-mass spectrometry. *J. Chinese Chem. Soc*. 47, 1155-1160.

Drewes, J.E., Bellona, C., Oedekoven, M., Xu, P., Kim, T., and Amy, G. (2005) Rejection of wastewater-derived micropollutants in high-pressure membranes leading to indirect potable reuse. *Environ. Progress* 24(4), 400-409.

Drewes, J.E. and Shore, L.S. (2001) Concerns about pharmaceuticals in water reuse, groundwater recharge, and animal waste, in *Pharmaceuticals and Personal Care Products in the Environment: Scientific and Regulatory Issues* (Daughton, C.G. and Jones-Lepp, T., Eds.), *Symposium Series 791*; American Chemical Society: Washington, D.C.

Dussault, E.B., Sherry, J.P., Lee, H.B., Burnison, B.K., Bennie, D.T., and Servos, M.R. (2005) In vivo estrogenicity of nonylphenol and its ethoxylates in the Canadian environment. *Hum. Ecol. Risk Assess*. 11(2), 353-364.

EDSTAC (Endocrine Disruptor Screening and Testing Advisory Committee) (1998) Final Report. (http://www.epa.gov/scipoly/oscpendo/history/finalrpt.htm)

Fang, H., Tong, W., Perkins, R., Soto, A.M., Prechtl, N.V., and Sheehan, D.M. (2000) Quantitative Comparisons of in Vitro Assays for Estrogenic Activities. *Environ Health Perspect*. 108(8), 723-729.

Fenske, M., Maack, G., Schafers, C., and Segner, H. (2005) An environmentally relevant concentration of estrogen induces arrest of male gonad development in zebrafish, *Danio rerio*. *Environ. Toxicol. Chem.* 24(5) 1088–1098.

Folmar, L.C., Denslow, N.D., Rao, V., Chow, M., Crain, D.A., Enblom, J., Marcino, J., and Guillette, L.J. Jr. (1996) Vitellogenin induction and reduced serum testosterone concentrations in feral male carp (*Cyprinus carpio*) captured near a major metropolitan sewage treatment plant. *Environ. Health Perspect.* 104(10), 1096-1101.

Folmar, L.C., Hemmer, M.J., Denslow, N.D., Kroll, K., Chen, J., Cheek, A., Richman, H., Meredith, H., and Gray, E.G. (2002) A comparison of the estrogenic potencies of estradiol, ethynylestradiol, diethylstilbestrol, nonylphenol and methoxychlor in vivo and in vitro. *Aquatic Toxicol.* 60, 101-110.

Furuichi, R., Kannan, K., Giesy, J.P., and Masunaga, S. (2004) Contribution of known endocrine disrupting substances to the estrongeic activity in Tama River water samples from Japan using instrumental analysis and in vitro reporter gene assay. *Water Research* 38(20), 4491-4501.

Gaido, K.W., Leonard, L.S., Lovell, S., Gould, J.C., Babaï, D., Portier, C.J., and McDonnell, D. P. (1997) Evaluations of chemicals with endocrine modulating activity in a yeast-based steroid hormone receptor gene transcription assay. *Toxicol. Appl. Pharm.* 143, 205-212.

Gillesby, B.E. and Zacharewski, T.R. (1998) Exoestrogens: mechanisms of action and strategies for identification and assessment. *Environ. Toxicol. Chem.* 17(1), 3-14.

Gray, J. and Sedlak, D. (2003) Removal of 17β-estradiol and 17α-ethinylestradiol in Engineered Treatment Wetlands. 3rd International Conference on Pharmaceuticals and Endocrine Disrupting Chemicals in Water. Minneapolis, Minn. March 19-21, 2003.

Gray, L.E. Jr., Kelce, W.R., Wiese, T., et al., (1997) Endocrine screening methods workshop report: detection of estrogenic and androgenic hormonal and anithormonal activity for chemicals that act via receptor or steroidogenic enzyme mechanisms. *Reproduct. Toxicol.* 11(5) 719-750.

Greenspan, F.S. and Strewler, G.J. (1997) *Basic and Clinical Endocrinology* Fifth Edition, Appleton and Lange, Los Altos, Calif.

Harries, J.D., Sheahan, D.A., Jobling, S., Matthiessen, P., Neall, M., Sumpter, J.P., Taylor, T., and Zaman, N. (1997) Estrogenic activity in five United Kingdom rivers detected by measurement of vitellogenesis in caged male trout. *Environ. Toxicol. Chem.* 16(3), 534-542.

Harris, C.A., Henttu, P. Parker, M.G., and Sumpter, J.P. (1997) The estrogenic activity of phthalate esters in vitro. *Environ. Health Perspect.* 105(8) 802-811.

Hemming, J., Barman, M., and Standridge, J. (2003) Assessment of Waters for Estrogenic Activity. Final Report. American Water Works Association Research Foundation. Denver, Colo.

Holbrook, R.D., Novak, J.T., Love, N.G., and Grizzard, T.J. (2001) Process Considerations for the Reduction of Endocrine Disruption Potential at Wastewater Treatment Facilities. WEFTEC 2001.

Holbrook, R.D., Novak, J.T., Grizzard, T.J., and Love, N.G. (2002) Estrogen receptor agonist fate during wastewater and biosolids treatment processes: a mass balance analysis. *Environ. Sci. Technol.* 36, 4533-4539.

Holbrook, R.D., Novak, J.T., and Love, N.G. (2005) Impact of activated sludge-derived colloidal organic carbon on behavior of estrogenic agonist recombinant yeast bioassay. *Environ. Toxicol. Chem.* 24(11), 2717-2724.

Howdeshell, K.L. (2002) A model of the development of the brain as a construct of the thyroid system. *Environ. Health Perspect.* 110(suppl. 3), 337-348.

Huang, C., and Sedlak, D. (2001) Analysis of estrogenic hormones in municipal wastewater effluent and surface water using enzyme-linked Immunosorbent assay and gas chromatography/tandem mass spectrometry. *Environ. Toxicol. Chem.* 20(1), 133-139.

Huggett, D.B., Foran, C.M., Brooks, B.W., Weston, J., Peterson, B., Marsh, K.E., La Point, T.W., and Schlenk, D. (2003) Comparison of in vitro and in vivo bioassays for estrogenicity in effluent from North American municipal wastewater facilities. *Toxicol. Sci.* 72, 77-83.

Ike, M., Jin, C.S, and Fujita, M. (2000) Biodegradation of bisphenol A in the aquatic environment. *Water Sci. Technol.* 42(7-8), 31-38.

Iwasaki, S., Kato, S., Takahashi, M., Kimura, T., Sakka, K., Ohmiya, K., Matsuda, T., and Matsui, S. (2004) Analysis of estrogen-like compounds in the environment by high performance liquid chromatography bioassay. *J. Biosci. Bioengineer.* 97(3), 216-218.

Jobling, S., Nolan, M., Tyler, C.R., Brighty, G., and Sumpter, J.P. (1998) Widespread sexual disruption in wild fish. *Environ. Sci. Technol.* 32, 2498-2506.

Jobling, S., Sheahan, D., Osborne, J.A., Matthiessen, P., and Sumpter, J.P. (1996) Inhibition of testicular growth in rainbow trout (*Oncorhynchus mykiss*) exposed to estrogenic alkylphenolic chemicals. *Environ. Toxicol. Chem.* 15(2), 194-202.

Jobling, S. and Sumpter, J.P. (1993) Detergent components in sewage effluent are weakly estrogenic to fish: an in vitro study using rainbow trout (*Onchorynchus mykiss*) hepatocytes. *Aquat. Toxicol.* 27, 361-372.

Johnson, A. and Sumpter, J. (2001) Removal of endocrine-disrupting chemicals in activated sludge treatment works. *Environ. Sci. Technol.* 35, 4697-4703.

Joss, A., Andersen, H., Ternes, T., Richle, P.R., and Siegrist, H. (2004) Removal of estrogens in municipal wastewater treatment under aerobic and anaerobic conditions: consequences for plant optimization. *Environ. Sci. Technol.* 38, 3047-3055.

Kavanagh, R.J., Balxh, G.C., Kiparissis, Y., Niimi, A.J., Sherry, J., Tinson, C., and Metcalfe, C.D. (2004) Endocrine disruption and altered gonadal development in white perch (*Morone Americana*) from the lower Great Lakes region. *Environ. Health Perspect.* 112, 898-902.

Kirk, L.A., Tyler, C.R., Lye, C.M., and Sumpter, J.P. (2002) Changes in estrogenic and androgenic activities at different stages of treatment in wastewater treatment works. *Environ. Toxicol. Chem.* 21(5), 972-979.

Kirschner, E.M. (1996) Growth of top 50 chemicals slowed in 1995 from very high 1994 rate. *Chem. Engineer. News* 74(15), 16-22.

Kitamura, S. Jinno, N., Ohta, S., Kuroki, H., and Fujimoto, N. (2002) Thyroid hormonal activity of the flame retardants tetrabromobisphenol A and tetrachlorobisphenol. *Biochem. Biophys. Res.* 26, 554-559.

Kreuzinger, M., Clara, M., Strenn, B., and Kroiss, H. (2004) Relevance of the sludge retention time (SRT) as design criteria for wastewater treatment plants for the removal of endocrine disruptors and pharmaceuticals from wastewater. *Water Sci. and Technol.* 50, 149-156.

Kuch, H.M. and Ballschmiter, K. (2001) Determination of endocrine-disrupting phenolic compounds and estrogens in surface and drinking water by HRGC-(NCI)-MS in the pictogram per liter range. *Environ. Sci. Technol.* 35, 3201-3206.

Langford, K.H., Scrimshaw, M.D., Birkett, J.W., and Lester, J.N. (2005) Degradation of nonylphenolic surfactants in activated sludge batch tests. *Water Research* 39,870-876.

Larsson, D.G.J., Hällman, H., and Förlin, L. (2000) More male fish embryos near a pulp mill. *Environ. Toxicol. Chem.* 19(12), 2911-2917.

Lauritzen, C. (1999) History of Estrogen Research, in *Estrogens and Antiestrogens I* (Oettel, M and Schillinger, E. Eds.) 1-14, Springer-Verlag, Berlin.

Layton, A., Gregory, B., Seward, J., Schultz, T., and Sayler, G. (2000) Mineralization of steroidal hormones by biosolids in wastewater treatment systems in Tennessee U.S.A. *Environ. Sci. Technol.* 34, 3925-3931.

Lee, H-B. and Peart, T.E. (2000) Bisphenol A contamination in Canadian municipal and industrial wastewater and sludge samples. *Water Qual. Res. J. Canada* 35(2), 283-298.

Lee, H-B., Peart, T.E., Chan, J., and Gris, G. (2004) Occurrence of endocrine-disrupting chemicals in sewage and sludge samples in Toronto, Canada. *Water Qual. Res. J. Canada* 39(1), 57-63.

Lee, Y.M., Oleszkiewicz, J.A., and Cicek, N. (2003) Fate of Pharmaceuticals and Endocrine Disrupting Compounds (EDC) in Municipal Wastewater Treatment Facilities: Methodology for Assessment of Environmental Levels. 3rd International Conference on Pharmaceuticals and Endocrine Disrupting Chemicals in Water. Minneapolis, Minn. March 19-21, 2003.

Lippert, T.H., Seeger, H., and Mueck, A.O. (1999) Metabolism of Endogenous Estrogens, in *Estrogens and Antiestrogens I* (Oettel, M and Schillinger, E. Eds.) 241-271, Springer-Verlag, Berlin.

Majewski, A.R., Blanchfield, P.J., Palace, V.P. and Wautier, K (2002) Waterborne 17α-Ethynylestradiol affects aggressive behaviour of male fathead minnows *(Pimephales promelas)* under artificial spawning conditions. *Water Qual. Res. J. Canada* 37(4), 697-710.

Mansell, J., Drewes, J.E., Rauch, T. (2004) Removal mechanism of endocrine disrupting compounds (steroids) during soil aquifer treatment. *Water Sci Technol.* 50, 229-237.

Matsui, S., Takigami, H., Matsuda, T., Taniguchi, N., Adachi, J., Kawami, H., and Shimizu, Y. (2000) Estrogen and estrogen mimics contamination in water and the role of sewage treatment. *Water Sci. Technol.* 42(12), 173-179.

Metcalf & Eddy, Inc. (2003) *Wastewater engineering: treatment and reuse* 4th ed. revised by Tchobanoglous, G., Burton, F.L., Stensel, H.D. Boston: McGraw-Hill.

Metcalfe, C.D., Metcalfe, T.L., Kiparissis, Y., Koenig, B.G., Khan, C., Hughes, R.J., Croley, T.R., March, R.E., and Potter, T. (2001) Estrogenic potency of chemicals detected in sewage treatment plant effluents as determined by in vivo assays with japanese medaka (*Oryzias latipes*). *Environ. Toxicol. Chem.* 20(2), 297-308.

Montgomery-Brown, J., Drewes, J.E., Fox, P., and Reinhard, M. (2003) Behavior of alkylphenol polyethoxylate metabolites during soil aquifer treatment. *Water Research* 37, 3672-3681.

Murk, A.J., Legler, J., van Lipzig, M.M.H., Meerman, J.H.N., Belfroid, A.C., Spenkelink, A., van der Burg, B., Rijs, G.B.J., and Vethaak, D. (2002) Detection of estrogenic potency in wastewater and surface water with three in vitro bioassays. *Environ. Toxicol. Chem.* 21(1), 16-23.

Nagel, S.C., vom Saal, F., Thayer, R.A., Dhar, M.G., Boechler, M., and Welshons, W.V. (1997) Relative binding affinity-serum modified access (RBA-SMA) assay predicts the relative in vivo bioactivity of the xenoestrogens bisphenol A and octylphenol. *Environ. Health Perspect.* 105, 70-76.

Nasu, M., Goto, M., Kato, H., Oshima, Y., Tanaka, H. (2001) Study on endocrine disrupting chemicals in wastewater treatment plants. *Water Sci Technol.* 43(2), 101-108.

National Research Council (1999) *Hormonally Active Agents in the Environment*. National Academy Press, Washington, D.C.

Nghiem, L.D., Schäfer A.I., and Waite, T.D. (2002) Adsorptive interactions between membranes and trace contaminants. *Desalination* 147, 269-274.

Nghiem, L.D., Schäfer, A.I., and Elimelech, M. (2004) Removal of natural hormones by nanofiltration membranes: measurement, modeling, and mechanisms. *Environ. Sci. Technol.* 38, 1888-1896.

Nieuwkoop, P.D. and Faber, F. (1994) Normal Table of *Xenopus laevis*. Garland, New York.

OECD (Organisation for Economic Co-operation and Development) (2004) Detailed Review Paper on Amphibian Metamorphosis Assay for the Detection of Thyroid Active Substances. OECD Series on Testing and Assessment No. 46. Paris.

OECD (Organisation for Economic Co-operation and Development) (2005) Draft Detailed Review Paper on Thyroid Hormone Disruption Assays. OECD Series on Testing and Assessment No. XX. Paris.

Onda, K., Nakamur, Y., Takatoh, C., Miya, A., and Katsu, Y. (2003) The behavior of estrogenic substances in the biological treatment process of sewage. *Water Sci. Technol*. 47(9), 109-116.

Orlando, E.F., Kolok, A. S., Binzcik, G.A., Gates, J.L., Horton, M.K., Lambright, C.S., Gray, L.E. Jr., Soto, A.M., and Guillette, L.J. Jr. (2004) Endocrine-disrupting effects of cattle feedlot effluent on an aquatic sentinel species, the fathead minnow. *Environ. Health Perspect*. 112(3), 353-358.3.

Panter, G.H., Thompson, R.S., and Sumpter, J.P. (1998) Adverse reproductive effects in male fathead minnows (*Pimephales promelas*) exposed to environmentally relevant concentrations of the natural oestrogens, oestradiol and oestrone. *Aquatic Toxicol*. 42, 243-253.

Purdom, C.E., Hardiman, P.A., Bye, V.J., Eno, N.C., Tyler, C.R. and Sumpter, J. P. (1994) Estrogenic effects of effluents from sewage treatment works. *Chem. Ecol.*8, 275-285.

Rodgers-Gray, T.P., Jobling, S., Morris, S., Kelly, C., Kirby, S., Janbakhsh, A., Harries, J.E., Waldock, M.J., Sumpter, J.P., and Tyler, C.R. (2000) Long-term temporal changes in the estrogenic composition of treated sewage effluent and its biological effects on fish. *Environ. Sci.Technol.* 34, 1521-1528.

Rosenfeldt E.J. and Linden, K.G. (2004) Degradation of endocrine disrupting chemicals bisphenol A, ethinyl estradiol, and estradiol during UV photolysis and advanced oxidation processes. *Environ. Sci. Technol.* 38, 5476-5483.

Routledge, E.J. and Sumpter, J.P. (1996) Estrogenic activity of surfactants and some of their degradation products assessed using a recombinant yeast screen. *Environ. Toxicol. Chem.* 15(3), 241-248.

Rudel, R.A., Melly, S.J., Geno, P.W., Sun, G., and Brody, J.G. (1998) Identification of alkylphenols and other estrogenic phenolic compounds in wastewater, septag, and groundwater on Cape Cod, Massachusetts. *Environ.. Sci. Technol.* 32(7), 861-869.

Seki, M., Yokota, H., Maeda, M., and Kobayashi, K. (2005) Fish full life-cycle testing for 17β-estradiol on medaka (*Oryzias latipes*). *Environ. Toxicol. Chem.* 24(5), 1259-1266.

Shi, J., Fujisawa, S., Nakai, S., and Hosomi, M. (2004) Biodegradation of natural and synthetic estrogens by nitrifying activated sludge and ammonia-oxidizing bacterium *Nitrosomonas europaea*. *Water Research*. 38, 2323-2330.

Shore, L., Gurevich, M., and Shemesh, M. (1993) Estrogen as an environmental pollutant. *Bull. Environ. Contam. and Toxicol.* 51, 361-366.

Snyder, S., Keith, T., Verbrugge, D., Snyder, E., Gross, T., Kannan, K., and Giesy, J. (1999) Analytical Methods for Detection of Selected Estrogenic Compounds in Aqueous Mixtures. *Environ. Sci. Technol.* 33, 2814-2820.

Snyder, S.A., Villeneuve, D.L., Snyder, E.M., and Giesy, J.P. (2001) Identification and quantification of estrogen receptor agonists in wastewater effluents. *Environ. Sci. Technol.* 35(18), 3620-3625.

Snyder, S., Vanderford, B., Adham, S., Oppenheimer, J., Gillogly, T., Bond, R., and Veerapaneni, V. (2004) Pilot Scale Evaluations of Membranes for the Removal of Endocrine Disruptors and Pharmaceuticals. AWWA Water Quality and Technology Conference. San Antonio, Texas November 14-18, 2004.

Solé, M., López de Alda, M. J., Castillo, M, Porte, C., Ladegaard-Pedersen, K., and Barceló, D. (2000) Estrogenicity determination in sewage treatment plants and surface waters from the Catalonian area (NE Spain). *Environ. Sci. Technol.* 34, 5076-5083.

Soto, A.M., Calabro, J.M., Prechtl, N.V., Yau, A.Y., Orlando, E.F., Daxenberger, A., Kolok, A. S., Guillette, L.J. Jr., le Bizec, B., Lange, I.G., and Sonnenschein, C. (2004) Androgenic and estrogenic activity in water bodies receiving cattle feedlot effluent in eastern Nebraska. *Environ. Health Perspect.* 112(3), 346-352.

Soto, A.M., Michaelson, C.L., Precthl, N.V., Weill, B.C., Sonnenschein, C., Olea-Serrano, F. and Olea, N. (1998) Assays to measure estrogen and androgen agonists and antagonists, in *In Vitro Germ Cell Developmental Toxicology: From Science to Social and Industrial Demand* (del Mao, J., Ed. 9-28, Plenum Press, New York.

Strenn, B., Clara, M., Gans, O., and Kreuzinger, N. (2003) Occurrence and Elimination of Selected Endocrine Disrupting Chemicals in Wastewater Treatment. 3rd International Conference on Pharmaceuticals and Endocrine Disrupting Chemicals in Water. Minneapolis, Minn. March 19-21, 2003.

Svenson, A. and Allard, A. (2004) Occurrence and some properties of the androgenic activity in municipal sewage effluents. *J. Environ. Sci. Health* A, 39(3), 693-701.

Svenson, A., Allard, A., and Ek, M. (2003) Removal of estrogenicity in Swedish municipal sewage treatment plants. *Water Research* 37, 4433-4443.

Szelei, J., Jimenez, J., Soto, A.M., Luizzi, M.F., and Sonnenschein, C. (1997) Androgen-induced inhibition of proliferation in human breast cancer MCF7 cells transfected with androgen receptor. *Endocrinol.* 138(4), 1406-1412.

Takigami, H., Taniguchi, N., Matsuda, T., Yamada, M., Shimizu, Y., and Matsui, S. (2000) The fate and behavior of human estrogens in a night soil treatment process. *Water Sci Technol.* 42, 45-51.

Tanaka, H., Yakou, Y., Takahaxhi, A., Higashitani, T., and Komori, K. (2001) Comparison between estrogenicities estimated from DNA recombinant yeast assay and from chemical analyses of endocrine disruptors during sewage treatment. *Water Sci. Technol.* 43(2), 125-132.

Tanghe, T., Devriese, G., and Verstraete, W. (1999) Nonylphenol and estrogenic activity in aquatic environmental samples. *J. Environ. Qual.* 28, 702-709.

Tashiro, Y., Takahira, K., and Nakanishi, Y. (2004) Improved application of recombinant yeast assays on environmental samples by size exclusion chromatography. *J. Environ. Monit.* 6, 546-551.

Ternes, T.A., Stumpf, M., Mueller, J., Haberer, K., Wilken, R.-D., and Servos, M. (1999a) Behavior and occurrence of estrogens in municipal sewage treatment plants –I. Investigations in Germany, Canada and Brazil. *Sci. Total Env.* 225, 81-90.

Ternes, T.A., Kreckel, P., and Mueller, J. (1999b) Behaviour and occurrence of estrogens in municipal sewage treatment plants – II. Aerobic batch experiments with activated sludge. *Sci. Total Environ.* 225, 91-99.

Thomas, K.V., Hurst, M.R., Matthiessen, P., McHugh, M., Smith, A., and Waldock, M.J. (2002) An assessment of in vitro androgenic activity and the identification of environmental androgens in United Kingdom estuaries. *Environ. Toxicol. Chem.* 21(7), 1456-1461.

Thorpe, K.L., Cummings, R.I., Hutchinson, T.H., Scholze, M., Brighty, G., Sumpter, J.P., and Tyler, C.R. (2003) Relative potencies and combination effects of steroidal estrogens in fish. *Environ. Sci. Technol.* 37, 1142-1149.

Tietge, J.E., Holcombe, G.W., Flynn, K.M., Kosian, P.A., Korte, J.J., Anderson, L.E. Wolf, D.C., and Degitz, S.J. (2005) Metamorphic inhibition of *Xenopus laevis* by sodium perchlorate: Effects on development and thyroid histology. *Environ. Toxicol. Chem.* 24(4), 926-933.

Van den Belt, K., Berheyen, R., and Witters, H. (2003) Comparison of vitellogenin responses in zebrafish and rainbow trout following exposure to environmental estrogens. *Ecotox. Environ. Safety* 56,271-281.

Vanderlaan, M. (1991) Immunoassays for Trace Chemical Analysis: Monitoring Toxic Chemicals in Humans, Food, and the Environment. American Chemical Society, Washington, D.C.

Vethaak, A.D., Lahr, J., Schrap, S.M., Belfroid, A.C., Rijs, G.B.J., Gerritsen, A., de Boer, J., Bulder, A.S., Grinwis, G.C.M., Kuiper, R.V., Legler, J., Murk, T.A.J., Peijnenburg, W., Verhaar,

J.M.H., and de Voogt, P. (2005) An integrated assessment of estrogenic contamination and biological effects in the aquatic environment of The Netherlands. *Chemosphere* 59, 511-524.

Vom Saal, F. (2000) Very Low Doses of Bisphenol A and Other Estrogenic Chemicals Alters Development in Mice. Proceedings Endocrine Disruptors and Pharmaceutical Active Compounds in Drinking Water. Workshop at the Center for Health Effects of Environmental Contamination, University of Iowa, April 19-21, 2000.

Weber, S., Leuschner, P., Kämpfer, P., Dott, W., and Hollender, J. (2005) Degradation of estradiol and ethinyl estradiol by activated sludge and by a defined mixed cluture. *Appl. Microbiol. Biotechnol.* 67, 106-112.

Wegener, G., Persin, J., Karrenbrock, F., Roerden, O., and Huebner, I. (1999) Presence and behavior of natural and synthetic estrogens and their conjugates in the aquatic environment. *Vom Wasser* 92, 347-360 (in German).

Westerhoff, P., Yoon, Y., Snyder, S., and Wert, E. (2005) Fate of endocrine-disruptor, pharmaceutical, and personal care product chemicals during simulated drinking water treatment processes. *Environ. Sci. Technol.* 39, 6649-6663.

Wintgens, T., Gallenkemper, M., and Melin, T. (2002) Endocrine disrupter removal from wastewater using membrane bioreactor and nanofiltration technology. *Desalination*. 146, 387-391.

Yamaoto, H. and Liljestrand, H.M. (2004) The fate of estrogenic compounds in the aquatic environment: sorption onto organic colloids. *Water Sci. Technol.* 47(9), 77-84.

Yoon, Y., Westerhoff, P., Snyder, S.A., and Esparza, M. (2003) HPLC-florescence detection and adsorption of bisphenol A 17β-estradiol, and 17α-ethinylestradiol on powered activated carbon. *Water Research* 37, 3530-3537.

Zacharewski, T. (1997) In vitro bioassays for assessing estrogenic substances. *Environ. Sci. Technol.* 31(3) 613-623.

## WASTEWATER UTILITY

**Alabama**
Montgomery Water Works & Sanitary Sewer Board

**Alaska**
Anchorage Water & Wastewater Utility

**Arizona**
Glendale, City of, Utilities Department
Mesa, City of
Peoria, City of
Phoenix Water Services Department
Pima County Wastewater Management
Safford, City of

**Arkansas**
Little Rock Wastewater Utility

**California**
Calaveras County Water District
Central Contra Costa Sanitary District
Corona, City of
Crestline Sanitation District
Delta Diablo Sanitation District
Dublin San Ramon Services District
East Bay Dischargers Authority
East Bay Municipal Utility District
Eastern Municipal Water District
El Dorado Irrigation District
Fairfield-Suisun Sewer District
Fresno Department of Public Utilities
Inland Empire Utilities Agency
Irvine Ranch Water District
Las Virgenes Municipal Water District
Livermore, City of
Los Angeles, City of
Los Angeles County, Sanitation Districts of
Napa Sanitation District
Orange County Sanitation District
Palo Alto, City of
Riverside, City of
Sacramento Regional County Sanitation District
San Diego Metropolitan Wastewater Department, City of
San Francisco, City & County of
San Jose, City of
Santa Barbara, City of
Santa Cruz, City of
Santa Rosa, City of
South Bayside System Authority
South Coast Water District
South Orange County Wastewater Authority
Stege Sanitary District
Sunnyvale, City of
Union Sanitary District
West Valley Sanitation District

**Colorado**
Aurora, City of
Boulder, City of
Greeley, City of
Littleton/Englewood Water Pollution Control Plant
Metro Wastewater Reclamation District, Denver

**Connecticut**
Greater New Haven WPCA

**District of Columbia**
District of Columbia Water & Sewer Authority

**Florida**
Broward, County of
Fort Lauderdale, City of
Miami-Dade Water & Sewer Authority
Orange County Utilities Department
Reedy Creek Improvement District
Seminole County Environmental Services
St. Petersburg, City of
Stuart Public Utilities
Tallahassee, City of
Tampa, City of
Toho Water Authority
West Palm Beach, City of

**Georgia**
Atlanta Department of Watershed Management
Augusta, City of
Clayton County Water Authority
Cobb County Water System
Columbus Water Works
Fulton County
Gwinnett County Department of Public Utilities
Savannah, City of

**Hawaii**
Honolulu, City & County of

**Idaho**
Boise, City of

**Illinois**
American Bottoms Wastewater Treatment Plant
Greater Peoria Sanitary District
Kankakee River Metropolitan Agency
Metropolitan Water Reclamation District of Greater Chicago
Wheaton Sanitary District

**Iowa**
Ames, City of
Cedar Rapids Wastewater Facility
Des Moines, City of
Iowa City

**Kansas**
Johnson County Unified Wastewater Districts
Unified Government of Wyandotte County/ Kansas City, City of

**Kentucky**
Louisville & Jefferson County Metropolitan Sewer District

**Louisiana**
Sewerage & Water Board of New Orleans

**Maine**
Bangor, City of
Portland Water District

**Maryland**
Anne Arundel County Bureau of Utility Operations
Howard County Department of Public Works
Washington Suburban Sanitary Commission

**Massachusetts**
Boston Water & Sewer Commission
Upper Blackstone Water Pollution Abatement District

**Michigan**
Ann Arbor, City of
Detroit, City of
Holland Board of Public Works
Saginaw, City of
Wayne County Department of Environment
Wyoming, City of

**Minnesota**
Rochester, City of
Western Lake Superior Sanitary District

**Missouri**
Independence, City of
Kansas City Missouri Water Services Department
Little Blue Valley Sewer District
Metropolitan St. Louis Sewer District

**Nebraska**
Lincoln Wastewater System

**Nevada**
Henderson, City of
Reno, City of

**New Jersey**
Bergen County Utilities Authority
Ocean, County of
Passaic Valley Sewerage Commissioners

**New York**
New York City Department of Environmental Protection

**North Carolina**
Charlotte/Mecklenburg Utilities
Durham, City of
Metropolitan Sewerage District of Buncombe County
Orange Water & Sewer Authority

**Ohio**
Akron, City of
Butler County Department of Environmental Services
Columbus, City of
Metropolitan Sewer District of Greater Cincinnati
Northeast Ohio Regional Sewer District
Summit, County of

**Oklahoma**
Oklahoma City Water & Wastewater Utility Department
Tulsa, City of

**Oregon**
Clean Water Services
Eugene/Springfield Water Pollution Control
Gresham, City of
Water Environment Services

**Pennsylvania**
Philadelphia, City of
University Area Joint Authority

**South Carolina**
Charleston Commissioners of Public Works
Mount Pleasant Waterworks & Sewer Commission
Spartanburg Sanitary Sewer District

**Tennessee**
Cleveland, City of
Knoxville Utilities Board
Murfreesboro Water & Sewer Department
Nashville Metro Water Services

**Texas**
Austin, City of
Dallas Water Utilities
Denton, City of
El Paso Water Utilities
Fort Worth, City of
Gulf Coast Waste Disposal Authority
Houston, City of
San Antonio Water System
Trinity River Authority

**Utah**
Salt Lake City Corporation

**Virginia**
Alexandria Sanitation Authority

Arlington, County of
Fairfax County
Hampton Roads Sanitation District
Henrico, County of
Hopewell Regional Wastewater Treatment Facility
Loudoun County Sanitation Authority
Lynchburg Regional WWTP
Prince William County Service Authority
Richmond, City of
Rivanna Water & Sewer Authority

**Washington**
Everett, City of
King County Department of Natural Resources
Seattle Public Utilities
Sunnyside, Port of
Yakima, City of

**Wisconsin**
Green Bay Metro Sewerage District
Kenosha Water Utility
Madison Metropolitan Sewerage District
Milwaukee Metropolitan Sewerage District
Racine, City of
Sheboygan Regional Wastewater Treatment
Wausau Water Works

**Australia**
South Australian Water Corporation
Sydney Water Corporation
Water Corporation of Western Australia

**Canada**
Greater Vancouver Regional District
Regina, City of, Saskatchewan
Toronto, City of, Ontario
Winnipeg, City of, Manitoba

**New Zealand**
Watercare Services Limited

**United Kingdom**
Yorkshire Water Services Limited

## STORMWATER UTILITY

**California**
Los Angeles, City of, Department of Public Works
Monterey, City of
Sacramento, County of
San Francisco, City & County of
Santa Rosa, City of
Sunnyvale, City of

**Colorado**
Aurora, City of
Boulder, City of

**Georgia**
Griffin, City of

**Iowa**
Cedar Rapids Wastewater Facility
Des Moines, City of

**Kansas**
Overland Park, City of

**Maine**
Portland Water District

**Minnesota**
Western Lake Superior Sanitary District

**North Carolina**
Charlotte, City of, Stormwater Services

**Pennsylvania**
Philadelphia, City of

**Tennessee**
Chattanooga Stormwater Management

**Texas**
Harris County Flood Control District, Texas

**Washington**
Bellevue Utilities Department
Seattle Public Utilities

## STATE

Arkansas Department of Environmental Quality
Connecticut Department of Environmental Protection
Fresno Metropolitan Flood Control District
Kansas Department of Health & Environment
Kentucky Department of Environmental Protection
Ohio River Valley Sanitation Commission
Urban Drainage & Flood Control District, CO

## CORPORATE

ADS Environmental Services
Alan Plummer & Associates
Alden Research Laboratory Inc.
Alpine Technology Inc.
Aqua-Aerobic Systems Inc.
Aquateam–Norwegian Water Technology Centre A/S
ARCADIS
Associated Engineering
Black & Veatch
Boyle Engineering Corporation
Brown & Caldwell
Burgess & Niple, Ltd.
Burns & McDonnell
CABE Associates Inc.
The Cadmus Group
Camp Dresser & McKee Inc.
Carollo Engineers Inc.
Carpenter Environmental Associates Inc.
CDS Technologies Inc.
CET Engineering Services
Chemtrac Systems Inc.
CH2M HILL
Conestoga-Rovers & Associates
CONTECH Stormwater Solutions
D&B/Guarino Engineers, LLC
Damon S. Williams Associates, LLC
Earth Tech Inc.
Ecovation
EMA Inc.
Environ/The ADVENT Group, Inc.
Fay, Spofford, & Thorndike Inc.
Freese & Nichols Inc.
ftn Associates Inc.
Fuss & O'Neill Inc.
Gannett Fleming Inc.
Geosyntec Consultants
GHD
Golder Associates Ltd.
Greeley and Hansen LLC
Hazen & Sawyer, P.C.
HDR Engineering Inc.
HNTB Corporation
Hydromantis Inc.
HydroQual Inc.
Infilco Degremont Inc.
Jacobson Helgoth Consultants Inc.
Jason Consultants LLC Inc.
Jordan, Jones, & Goulding Inc.
KCI Technologies Inc.
Kelly & Weaver, P.C.
Kennedy/Jenks Consultants
KMK Consultants
Komline Sanderson Engineering Corporation
Limno-Tech Inc.
Lombardo Associates Inc.
The Low Impact Development Center Inc.
Malcolm Pirnie Inc.
Material Matters
McKim & Creed
MEC Analytical Systems, Inc.
Metcalf & Eddy Inc.
Monteco Corporation
MPR Engineering Corporation, Inc.
MWH
O'Brien & Gere Engineers Inc.
Odor & Corrosion Technology Consultants Inc.
Oscar Larson & Associates
PA Government Services Inc.
Parametrix Inc.
Parsons
Post, Buckley, Schuh & Jernigan
RMC Water & Environment
R.M. Towill Corporation
Ross & Associates Ltd.
Rothberg, Tamburini & Windsor, Inc.
SAIC
Siemens Water Technologies
Stantec Consulting Inc.
Stearns & Wheler, LLC
Stone Environmental Inc.
Stratus Consulting Inc.
Synagro Technologies Inc.
Tetra Tech Inc.
Trojan Technologies Inc.
Trussell Technologies, Inc.
URS Corporation
Wade-Trim Inc.
Westin Engineering Inc.
Weston Solutions Inc.
Woodard & Curran
Zenon Environmental Inc.
Zoeller Pump Company

## INDUSTRY

American Electric Power
American Water
ChevronTexaco Energy Research & Technology Company
The Coca-Cola Company
Dow Chemical Company
DuPont Company
Eastman Chemical Company
Eastman Kodak Company
Eli Lilly & Company
Merck & Company Inc.
Premier Chemicals LLC
Procter & Gamble Company
RWE Thames Water Plc
Severn Trent Services Inc.
Suez Environment
United Water Services LLC

**Note: List as of 9/1/06**